CRYOGENIC
LABORATORY EQUIPMENT

THE INTERNATIONAL CRYOGENICS MONOGRAPH SERIES

CRYOGENIC
LABORATORY EQUIPMENT

A. J. Croft
Clarendon Laboratory, University of Oxford

PLENUM PRESS • NEW YORK–LONDON • 1970

Library of Congress Catalog Card Number 65-11337
SBN 306-30253-5

Dedicated with respect and affection to the memory of
FRANZ EUGEN SIMON
with gratitude for the risk which he took in 1947

Preface

This book is meant for laboratory workers who for one reason or another have a need to cool something down to temperatures below that of liquid nitrogen — notably to $4.2°K$ and below. It does not deal with experimental techniques at low temperatures, but I have tried to bring the reader face to face with the brutish realities of the necessary hardware. As well as giving information about sources of supply of equipment, I have gone into some detail about how some of it can be made in laboratory workshops for the sake of those who are short of money but blessed with competent technical support. So far as highly specialized items such as liquefiers, refrigerators, refrigerant containers, cryostat dewars, etc., are concerned, I have included all sources of supply which I have got to hear of; in the case of more generally available equipment only representative sources of known reliability have been quoted. Any omissions or errors must be put down either to my own ignorance, stupidity, or lack of will to get about the world, or perhaps to the difficulty I have had in extracting information from manufacturers. However, most have gone to great trouble to help, and I hope I have done them justice.

Brought up to work indifferently in inches and centimetres and perched between the opposing pulls of the USA and Europe, I have used a mixture of units which may shock the purist. Manufacturers' information is given in the units quoted by them, and nonmetric units will be found where equipment and materials are commonly available only in such units at the time of writing. However, the calorie, the British thermal unit, and the degrees Fahrenheit and Rankine have been banished.

I am grateful to Dr. Kurt Mendelssohn for suggesting that I should write this book, and to the publishers for their patience and generosity. Too many people have helped with its preparation for me to be able to name them all, but I must record my gratitude to Dr. Peter McClintock (now at the University of Lancaster) and Mr. Michael Wells of the Clarendon Laboratory. Prof. Nicholas Kurti has been keeping my thermodynamics on the rails for many years, and but for him this book would include more loose statements than it does. In the matter of getting the text presentable, I offer my warm thanks to my wife for her fortitude as amanuensis to one overendowed with flexibility of purpose, and to Mrs. Patricia Band for her interpretative skill in producing an accurate typescript from a consequently chaotic manuscript.

Clarendon Laboratory   A. J. Croft
University of Oxford
October 1968

Contents

CRYOGENIC
LABORATORY EQUIPMENT

Chapter 1

Liquid Refrigerants

1.1. LIQUID REFRIGERANTS IN GENERAL

The fundamental requirement for cryogenic work is a means of absorbing heat at a low temperature. This can be achieved in two distinct ways: a bath of liquid can be used which takes up heat as heat of evaporation (or in the case of subcooled liquid by rise of temperature), or heat can be transferred to a fluid medium — usually a gas — which is refrigerated in a continuous cycle. Sometimes it is convenient to combine both types of system: a quantity of liquid, for instance, can be used as a buffer for a refrigeration system.

A refrigerator may be the most economical means of providing a single heat sink, whereas in a laboratory where many experiments are done at low temperatures a multiplicity of refrigerators would be cumbersome and expensive, and a supply of appropriate liquid refrigerants more economical. Where a wide temperature range has to be covered a refrigerator may offer advantages over the use of liquid refrigerants. Refrigerators are discussed together with liquefiers in Chapter 3; here we discuss the properties and uses of the common liquid refrigerants.

A bath of a liquid refrigerant boiling at atmospheric pressure offers a convenient heat sink — provided that there is a substance having a suitable boiling point — the temperature of which stays constant to the extent that atmospheric pressure stays constant, and in any case this temperature can be accurately determined by measuring the pressure.(This is subject to the qualification made below in the case of liquid hydrogen). Cryogenics would be very different were atmospheric pressure to be 1% of what it is: only oxygen and the two isotopes of helium would be liquid at atmospheric pressure. The range over which liquid refrigerants can be used is that between the critical and triple point temperatures shown in Table I. The use of solid refrigerants is not impossible, but

TABLE I

Some Physical Properties of Cryogenic Refrigerants

Refrigerant	Boiling point at 760 mm Hg (°K)	Density (g·cm⁻³)	Latent heat (J·g⁻¹)	Critical point		Triple point		Volume ratio at 15°C (gas/liquid)	Enthalpy change from boiling point to 280°K (J·g⁻¹)
				Temp. (°K)	Pressure (atm)	Temp. (°K)	Pressure (mm Hg)		
Oxygen	90.2	1.14	213	154.8	50.1	54.4	1.14	844	174
Nitrogen	77.3	0.81	199	126.2	33.5	63.1	96.4	681	217
Neon	27.2	1.20	89.8	44.5	25.9	24.6	324	1410	263
Hydrogen (n)	20.4	0.07	451	33.25	12.8	14.0	54.0	824	3428
Helium-4	4.18	0.125	20.5	5.25	2.26	—	—	742	1425
Helium-3	3.20	0.059	8.2	3.35	1.2	—	—	440	—

presents difficulties. In practice, the direct use of liquid or solid refrigerants is out of the question for temperatures in the ranges $\sim 5 - \sim 12°K$ and $\sim 30 - \sim 60°K$.

1.2. SPECIFIC LIQUID REFRIGERANTS

Liquid nitrogen is now not merely an everyday laboratory commodity, but an industrial raw material of importance. This is because the most economical way of separating the constituents of air is by distillation, and it is in the liquid form that bulk gases are most cheaply transported and stored. The minority who are more interested in the "cold"* than in the substance benefit from this technological fact. The scale on which liquid nitrogen is produced in industrially developed parts of the world is such that the cost may be low and the reliability of supply high. It will usually be the case therefore that liquid nitrogen will be bought, and stored in a suitable vessel (Chapter 2). Where supplies are not available or are expensive a commercial liquefier (Chapter 3) may be used. In borderline cases a cost analysis may be made on the lines suggested below.

In the cryogenic laboratory liquid nitrogen is used for the following purposes:

for maintaining an intermediate heat sink between room temperature and parts at lower temperatures, although with modern heat insulation techniques this is not always obligatory (Chapter 7);

for providing precooling in liquefiers and refrigerators (Chapter 3);

for precooling apparatus later to be used at a lower temperature, although where light systems are in use and efficient use of the cold of the evaporated vapour can be made this may not be necessary;

for cooling adsorbents used for purifying gases (Chapter 5);

Liquid nitrogen is cold enough to condense air from the atmosphere. This can cause errors in accurate temperature measurements and also unexpected condensations. These, in subsequently closed systems, may result in bursts. Apart from this, liquid nitrogen is essentially a safe substance, except for the physiological hazard mentioned in Section 9.3.

Liquid oxygen is not to be ignored as a refrigerant. It may be cheaper than liquid nitrogen and more readily available, although precautions have to be taken in its use and there are circumstances in which it should not be used at all (see Section 9.2).

Liquid air comes into the picture only in that it is the primary product of laboratory liquefiers: capital and running costs can be saved

---

*The concept of "cold" is a useful one, and the word will henceforward appear without the inverted commas.

if the rectification column needed for liquid-nitrogen production is left out. Paradoxically, the only way of getting liquid air commercially is to buy liquid nitrogen and liquid oxygen and then to mix them.

Liquid neon is coming into use as a safer alternative to liquid hydrogen for applications in which a temperature of 27°K is low enough, because neon gas is now less expensive than formerly. Although its use is still uncommon in laboratories, a liquid-neon refrigeration system capable of handling 1 MW has been built for cooling a solenoid having high-purity aluminium conductors at NASA Lewis Research Center, Cleveland, Ohio. Liquid neon has the added advantage of packing in over three times more latent heat per unit volume than liquid hydrogen. However, its triple point is at 24.6°K (324 mm Hg) and this makes it unsuitable for precooling in an internal-work helium liquefier.

Liquid hydrogen is now produced on a large scale as a rocket fuel: the plant at West Palm Beach, Florida, installed by Air Products and Chemicals, Inc., produces 28 British tons per day. In the laboratory the ready oxidizability of hydrogen has led to a general reluctance to use it. This is a pity because it is of great value in experimental work. The dangers are not recondite, and can be readily guarded against if properly faced up to (see Section 9.2). Indeed some chemical engineers say that liquid hydrogen is less dangerous to handle than liquid hydrocarbons in that evaporated hydrogen is lighter than air, whereas dangerous concentrations of evaporated hydrocarbons may lie about at ground level for considerable periods.

Liquid hydrogen may be readily produced in a Linde liquefier in which liquid nitrogen is used for precooling, or in other ways (Chapter 3). On the other hand, if one is within reach of a large hydrogen liquefaction plant it may be bought at a very low price. At present liquid hydrogen is produced commercially in Great Britain only on a small scale and at a relatively high cost. It is therefore the case that in the USA it is uneconomical to make one's own liquid hydrogen if one is within reach of a supplier, whereas in Great Britain those needing appreciable quantities are better off with their own liquefier.

There is a transition between two states of the hydrogen molecule, which is important because a relatively large amount of energy is involved. Hydrogen for which the spins of the two protons are parallel is called *ortho*-hydrogen; in *para*-hydrogen they are antiparallel. At very high temperatures statistical arguments show that out of every four molecules, three will be in the *ortho* and one in the *para* state. (The equilibrium composition at 300°K is 25.07% *para*-hydrogen. At low temperatures the equilibrium composition is almost all *para*-hydrogen — 99.82% at 20°K). At 20.4°K the transition in the liquid phase from the *ortho* to the *para* state occurs spontaneously (i.e., in the absence of a catalyst) with a time constant of 3.29 days and an evolution of heat of 669 J-g^{-1} (cf. 451 J-g^{-1} for the latent heat). [1] This phenomenon is of importance in the economics of liquefying hydrogen, since it both increases the consumption of liquid nitrogen and decreases the yield of the liquefier by +c.40% and −c.30% respectively for full conversion. In cases where liquid hydrogen has to be stored for periods exceeding ten days or where there are special reasons why boil-off due to conversion is unacceptable, full conversion is economic. Where liquid hydrogen is used within hours of its being made it is usually desirable to avoid conversion rather than to promote it. In intermediate cases it may be economic to convert to the $77°K$ equilibrium mixture, which is approximately 50/50. One should not lose sight of the different vapour pressures of the two forms at the same temperature (Table II); this can introduce errors into temperature measurements.

Some cryogenic work can be done at temperatures above $14°K$ — the triple point of hydrogen — but the principal laboratory uses of liquid hydrogen are as an adjunct to work with liquid helium. Its part in the liquefaction of helium will be discussed later; here we take note of its value as a precooling agent for work at liquid-helium temperature. (Readers lucky enough to have access to unlimited quantities of liquid helium should skip the next paragraph: it is intended for the poverty-stricken.)

Suppose that it is required to cool 1 kg of copper from $77°K$ to $4°K$ and that it is possible to do this in the following three idealized ways: (1) cooling by latent heat of liquid helium only; (2) cooling by latent heat of liquid helium but with the maximum possible use of the cold of the evaporated vapour; (3) cooling first by latent heat only of liquid hydrogen and then by latent heat only of liquid helium. The quantities of liquid hydrogen and liquid helium required in these three ideal cases are given in Table III.

The dramatic difference in the figures for the consumption of liquid helium in these three cases is accounted for by the small latent heat of liquid helium compared with the specific heat of the vapour and the fall in specific heat of copper with falling temperature. (See Table XV for data on this and other common constructional materials.)

The moral is that precooling with liquid hydrogen — or, alterna-

TABLE II

Vapour Pressures of *Para* and *Normal* Liquid Hydrogen

	Vapour pressure (mm Hg)			
	at $14°K$	at $16°K$	at $18°K$	at $29°K$
99.79% *para*-hydrogen	58.8	161.2	360.6	700.3
25.00% *para*-hydrogen	55.4	153.3	345.9	675.7

TABLE III

Case	Liquid hydrogen (cm^3)	Liquid helium (cm^3)
1	–	2300
2	–	140
3	190	13

tively, liquid neon – will reduce the quantity of liquid helium required very considerably. Where precooling is not possible the greatest use should be made of the cold of the evaporated helium, e.g., by inserting gauzes into a chamber to promote thermal exchange between the vapour and the wall of the chamber. In practice, of course, it is rarely possible to make use of all of the cold of the evaporated vapour, just as it is impossible to avoid making use of some of it; the figures given for cases 1 and 2 above represent extremes.

Liquid helium * is by far the commonest refrigerant for work at temperatures below that of liquid nitrogen. Until recently liquid helium was only produced for cryogenic use, but in the USA markets for helium gas in the East are supplied by means of a fleet of transcontinental 10,000-gal liquid-helium road tankers. This development has increased the availability of liquid helium and lowered the cost. There has therefore been a growing tendency to purchase liquid helium rather than to install a liquefier, although it is always worth making a cost analysis first (see below).

Except where only small quantities are in use, helium gas recovery is the rule outside the USA. In cases where economics are on the side of making liquid helium rather than buying it, a convenient way of operating is to buy make-up helium in the form of liquid, since there is then no need to man-handle gas cylinders.

Liquid helium-3 is now used in modest quantities for experimental work in closed-condensation or continuous-refrigeration systems. The light isotope is present in atmospheric and in mineral helium in very small concentrations – 1 and 0.1 ppm, respectively – but it is now available as a product of the radioactive decay of tritium (half-life 12.5 years) at prices varying according to purity. Its vapour pressure is compared with that of the isotope of mass 4 in Table IV and unlike the latter it does not show superfluidity. The phase separation of mixtures of the two isotopes, on which a continuous refrigeration cycle has been based, is discussed in Section 3.5.

---

*Unless otherwise specified, "helium" refers to the isotope of mass 4.

1.3. REFRIGERANT ECONOMICS

In parts of the world where refrigerants can be bought the choice may have to be made between buying or making one's own; in some cases the former alternative may be so obviously the better that no comparative cost analysis needs making. The position is more complicated where gas recovery has to be considered, and we shall therefore think in terms of liquid helium, although parts of the discussion are relevant to other refrigerants.

When working out the real cost of buying liquid helium one must take into account the credit given for recovered gas less the cost of recovering it. Where substantial quantities are in use the cost of recovery equipment can be justified, but one should not lose sight of the fact that the standard helium cylinder weighing 140 lb contains the equivalent of only 10 litres of liquid and the cost of handling many cylinders within the laboratory may be considerable.

In a commercial organization the cost of producing liquid helium can be worked out according to principles which are a matter of everyday business practice. However, in a university laboratory there may be a distinction between capital and running costs and overheads which may tip the balance in favour of owning a liquefier. To take an extreme case: a laboratory may have access to funds for purchasing equipment but may be short of funds for recurrent expenditure, it may have space available and thereby incur no additional costs for housing a liquefier, and it may have staff who can be deployed on running one — especially if it is automatic — and on maintaining it. The additional cost to the laboratory of producing liquid helium is then limited to that of liquid nitrogen (if required), make-up helium gas, electrical power, cooling water, and spare parts and other materials for maintenance. If costing on such a basis is appropriate, it may prove much cheaper to make one's own liquid helium than to buy it. In any

TABLE IV

Vapour Pressures of Liquid ^{3}He and ^{4}He

Temp. ($^\circ$K)	Vapour pressures (mm Hg)	
	^{3}He	^{4}He
1	8.56	0.12
0.6	0.50	2.8×10^{-4}
0.3	0.0015	3×10^{-10}

specific case the basis for working out the real cost will lie between the extreme lower limit just indicated and the upper limit in which depreciation, interest charges, labour costs, and overheads are included.

REFERENCE

1. Larsen, A.H., Simon, F.E., and Swenson, C.A., *Rev. Sci. Instr.* 19, 266 (1948); see also Farkas, A., *Ortho-Para Hydrogen and Heavy Hydrogen*, Cambridge University Press, London and New York (1935).

Chapter 2

Storage and Handling of Liquid Refrigerants

2.1 STORAGE VESSELS – PRINCIPLES

The extent of the problem of keeping heat out of a vessel containing a liquid refrigerant varies widely. Before we review the design of actual vessels it will put matters on a quantitative basis to consider two extreme cases, both of which resemble those arising in practice: a 5000-litre liquid-nitrogen tank and a 100-litre liquid-helium vessel.

First we must decide on the permissible heat leaks: insulation costs money, and an economic balance must be struck. In the case of the liquid-nitrogen tank a loss of 1% per day is acceptable – this corresponds to a total heat influx of about 100 W. A 1% per day loss from the 100-litre liquid-helium vessel demands the much smaller heat influx of 30 mW.

Taking the more difficult case first, let us consider how heat can get into the idealized classical dewar vessel shown in Fig. 2.1. The three principal routes are (1) thermal conduction down the material of the neck tube, (2) radiation from the outer shell to the inner shell, and (3) conduction through the residual gas between the two shells. Thermal conduction through the gas in the neck tube and thermal radiation down it can usually be neglected.

In the case of a 100-litre liquid-helium vessel of this design the quantitative contributions of these sources of heat influx are as follows:

1. The neck tube will be made from a material having as low a thermal conductivity as possible and the design will be such that the tube has the smallest possible cross-sectional area and the greatest possible length consistent with mechanical considerations. Taking as an example a 5/8-in.-diam × 0.006-in.-wall-thickness stainless steel tube, 24 in. long, with the top at 15°C and the bottom at 4°K, we find a heat flow of 75 mW. If a thermal anchor to 77°K can be made at a point

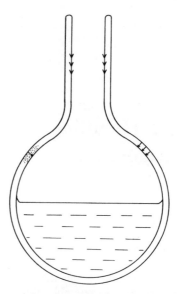

Fig. 2.1. Diagrammatic representation of
the basic dewar vessel showing sources
of heat influx.

one third of the way down, the heat flow reaching the bottom end at
4°K will be reduced to 13 mW. In practice, the heat flow down a neck
tube is considerably reduced by thermal exchange with the cold vapour
coming up from the liquid helium (see Wexler[1]). The use of standard
"cryogenic" tubing therefore provides an acceptable solution for the cases
we are considering; for lower heat influxes techniques involving multiple
coaxial tubes, corrugated tubes, or the use of plastics can be useful.

2. Where the emissivities of two surfaces are approximately equal
and much less than unity, and where the temperature of one is much
lower than that of the other — i.e., $T_1^4 \gg T_2^4$ — the heat transmitted
by thermal radiation can be expressed to a good approximation by

$$\dot{Q} = \frac{\sigma A T_1^4 e}{2} \text{ W}$$

where T_1 is the temperature in deg K of the hotter surface and σ
(Stefan's constant) has the value 5.67×10^{-12} W-cm^{-2} - (deg K)$^{-4}$.
Taking emissivity $e = 0.02$ — a representative minimum figure — we find
for the radiation from 15°C the unacceptable figure of 4.4 W. It is clear
that something must be done to reduce thermal radiation, and an
obvious step would be to insert a radiation shield cooled to 77°K by
liquid nitrogen. The reduction in radiation will then be $(288/77)^4 = 194$,
giving the acceptable figure of 23 mW. As will be seen below, there are

more sophisticated ways of baffling thermal radiation — notably "super-insulation."

3. White[2] gives an expression for heat conduction at low pressures which in the case of helium can be reduced to

$$Q \approx 9pT \quad \text{mW-cm}^{-2}$$

where p is expressed in mm Hg and T in deg K. At a pressure of 10^{-6} mm Hg this gives a heat influx of 29 mW for $T = 288°$K. However, it has been made clear above that the reduction of thermal radiation demands a surrounding surface at a lower temperature, and, moreover, the surface at 4°K will cryopump the vacuum space to a pressure lower than 10^{-6} mm Hg. The contribution of gaseous conduction in liquid-helium and liquid-hydrogen vessels is therefore negligible in the absence of leaks from the inner container.

Returning to the 5000-litre liquid-nitrogen vessel, we find that the contributions of these three sources of heat influx are very different. First we must note the effect of change of scale. For shapes likely to be used for tanks the surface area increases as the 2/3-power of the volume, i.e., the larger the vessel, the less important become the contributions related to surface area. Further, the conduction along pipes communicating with the outside has to increase only slightly with increase in volume once we are concerned with vessels holding 10^3 litres and more. In this case we are therefore little concerned with the conduction down connecting pipes, and radiation is of less, though still of some, importance. The principal problem arises from the difficulty of maintaining a low enough pressure in the space between the inner and outer walls to keep the gaseous conduction down to a low enough limit without the use of large and expensive high-vacuum pumps. The solution lies in the use of certain insulating powders (see Chapter 7) the pores of which are smaller than the mean free path of gases at pressures which can readily be maintained by the occasional use of a mechanical pump — i.e., about 10^{-3} mm Hg. The apparent mean thermal conductivity of such powders between room temperature and liquid-nitrogen temperature is ~ 10 μW-cm^{-1}-(deg K)$^{-1}$ at a pressure of 10^{-3} mm Hg. Application of this value to a hypothetical tank holding 5000 litres gives a heat influx of 20 W.

2.2 SMALL LIQUID-NITROGEN VESSELS

Multilayer insulation — commonly referred to as superinsulation — is treated in detail in Chapter 7. It has entirely revolutionized the design of cryogenic refrigerant vessels, including the 25-litre liquid-nitrogen vessel which has been ubiquitous in laboratories of all sorts for decades.

TABLE V

Vessel construction	Height (cm)	O.D. (cm)	Weight (kg)	Loss per day (%)
Traditional	75	46	29	3–4½%
Superinsulated	61	43	6	<4%

TABLE VI

Superinsulated Liquid-Nitrogen Vessels (litres)

Supplier*	25	50	100	200/250	500
L'Air Liquide(Ai_1)	X	X	X	–	X
British Oxygen Cryoproducts(Br)	X	–	–	X	X
Hofman Laboratories Inc.(Ho_1)	X	X	–	–	–
Spembley Technical Products(Sp)	X	–	–	–	–
Union Carbide (Un_1)	X	X	–	–	–

*The superscripts refer to the list of addresses given in the appendix, where sources of supply are also listed.

The traditional construction embodied two copper spherical shells with high-vacuum insulation between them, with the whole assembly contained in a sheet steel case. Its modern counterpart is made of aluminium with superinsulation. Comparative physical properties are shown in Table V. Although vessels of the traditional construction are still obtainable from many makers, Table VI refers only to superinsulated vessels.

For dispensing liquid nitrogen in small quantities in the laboratory the traditional glass dewar vessel remains the cheapest and most satisfactory. For bucket-sized vessels the fragility of glass becomes a disadvantage, and the cylindrical metal or plastic vessels offered by many suppliers are more durable, though much more expensive. (See the Appendix for suppliers.)

2.3. LARGE LIQUID-NITROGEN TANKS

Laboratories which buy their liquid nitrogen are likely to find that they are best suited by a tank which holds the contents of a road

TABLE VII

Large Liquid-Nitrogen Tanks (litres)

Supplier*	100	500	1000	5000	10,000
L'Air Liquide (Ai_1)		———————————			
Butterfield (Bu)			——————		
Hofman (Ho)			—————————————		
Ronan and Kunzel (Ro)			———————————————		
Union Carbide (Un_1) (superinsulated)		——————			

Fig. 2.2. 337-gal liquid-nitrogen tank made by W. P. Butterfield (Engineers) Ltd. [Photo: W.P. Butterfield (Engineers) Ltd.]

tanker – i.e., about 5000 litres. These are usually insulated with an expanded mineral powder such as Perlite (Chapter 7) in a space maintained at a pressure of $10^{-2}-10^{-3}$ mm Hg by the use of a mechanical vacuum pump every few months. Table VII indicates the range of sizes obtainable from various makers.

Figure 2.2 shows a 337-gal container made by W. P. Butterfield Ltd. The loss rate from this type of vessel varies from 0.5% for the larger vessels to 2.5% for the smaller, being lower, size for size, for superinsulated vessels. They are usually fitted with gravity-fed pressure build-up coils and hydrostatic level gauges and invariably with pressure-relief valves or bursting disks.

2.4. LIQUID-HELIUM AND LIQUID-HYDROGEN VESSELS

Since so far as laboratory work is concerned liquid helium is so much more widely used than liquid hydrogen, we shall be discussing vessels for the former only; these are of course suitable for the latter as well. However, where vessels are specifically required for liquid hydrogen some makers offer a vessel in which the degree of insulation is reduced so as to relate to the higher latent heat of liquid hydrogen; greater capacity and lower cost are then achieved for given outside dimensions.

For many years most liquid-helium storage vessels were constructed by enclosing one double-walled metal dewar vessel inside another. The space between the two vessels was filled with liquid nitrogen so that the surface radiating to the liquid-helium container was at 77°K, and the top of the neck tube of the inner vessel was tied to this temperature. These vessels have low evaporation rates – typically 0.5% per day for a 100-litre vessel. However, they suffer from the disadvantages of bulk, weight, fragility, and the necessity for liquid nitrogen. It will be seen below that their superinsulated successors have evaporation rates two to four times higher, but this is heavily outweighed by their many advantages, especially now that liquid helium has become cheaper. A typical example is the 100-litre vessel shown in Fig. 2.3. The Clarendon Laboratory uses 50-litre vessels made by British Oxygen Cryoproducts Ltd. (see Fig. 2.4), the dimensions and performance of which are as follows:

Overall height	44½ in.
Outer diameter	19½ in.
Inner diameter	14¾ in.
Neck-tube inner diameter	1¼ in.
Weight empty	165 lb
Evaporation rate	1.8% per day

Fig. 2.3. 100-litre superinsulated liquid-helium vessel made by
the Linde Division, Union Carbide Corp. (Photo: Union Carbide
Corp.)

TABLE VIII
Superinsulated Liquid-Helium Vessels (litres)

Supplier*	25	35	50	100	250	500	1000
L'Air Liquide(Ai_1)	X	–	–	X	–	–	–
BOC(Br_1)	–	–	X	X	X	X	X
Cryenco(Cr_3)	–	–	–	X	X	X	X
Gardner Cryogenics(Ga_1)	–	X	–	X	–	–	–
Union Carbide(GB)	–	X	–	X	–	–	–
Union Carbide(USA)	–	–	–	X	–	X	–

Vessels of this type are available as indicated in Table VIII at the time of writing, but other makers and other sizes are likely to appear on the market in the near future.

These vessels embody multilayer insulation – usually of the aluminium foil and glass fibre paper type – but with an important addition which greatly improves the performance. Layers of superinsulation are seperated by thin, highly-conducting copper sheets the upper ends of which are strapped to appropriate points on the outside of the neck tube. In this way some of the heat which would otherwise reach the liquid is transferred to the vapour passing up the neck, the cold of which would be wasted, except in so far as it reduces the thermal conduction down the neck tube. The number of these shields is commonly in the range 5–10.

2.5. THE HANDLING OF LIQUID-REFRIGERANT STORAGE VESSELS

In considering how to handle storage vessels for liquid refrigerants we must give top priority to the avoidance of internal overpressures. The manufacturer usually includes bursting disks, etc., which prevent disaster to the user in the event of an overpressure, but which may result in very serious damage to the vessel (see Chapter 9).

Superinsulated vessels require precooling with liquid nitrogen. Where these have inner vessels made of stainless steel it is worth making sure that the whole height of the vessel reaches the temperature of the liquid nitrogen. The nitrogen vapour should be replaced by helium gas, since otherwise it is possible for a plug of solid nitrogen to form in the neck when the vessel is filled with liquid helium. When a storage vessel is used to fill a large number of cryostats (e.g., when quantities of, say, a few litres are withdrawn from a 100-litre vessel) it is important to include a valve in the withdrawal transfer line so that the vessel can be kept at the same pressure, typically about 1 lb-in^{-2}. The alternative of cyclically letting the pressure down to atmospheric and then repressurizing leads to a considerable heat influx – the greater, the emptier is the vessel. The magnitude of this effect is partly due to the unusually small ratio of the density of liquid helium to that of its vapour at the boiling point, namely, 7.5.

This figure is also worth bearing in mind when a storage vessel has been emptied of its liquid contents; if the vapour is all at 4.2°K, a 100-litre vessel contains the equivalent in gas of about 13 litres of liquid helium, i.e., about 340 ft^3 at NTP.

The question of level measurement is dealt with in Chapter 6, but it is worth pointing out that the contents of a *superinsulated* vessel can readily be measured on a sufficiently sensitive weighing machine.

Fig. 2.4. Transfer of liquid helium into a cryostat from a British Oxygen Cryo-products 50-litre superinsulated vessel. (Photo: R. W. Bowl, Clarendon Laboratory.)

Some cryostats can be brought to a storage vessel to be filled, but usually it is the other way round. When planning a low-temperature laboratory from scratch one can fit each cryostat with a transfer line embodying half of a horizontal Johnston coupling (see below) at a standard height. This height will be determined by the combined heights of the trolleys, the storage vessels, and the transfer line terminating in the mating half of the Johnston coupling. Where such standardization is not possible one of the commercially available lifting trolleys can be used. Figure 2.4 shows liquid helium being transferred to a cryostat in the Clarendon Laboratory from a 50-litre superinsulated vessel. It will be seen that the Johnston coupling is vertical here, because there were a large number of cryostats fitted with transfer lines originally designed for small intermediary vessels. The operator is adjusting the valve which controls the flow of liquid helium through the transfer line. Note that this has been fabricated from standard copper tubing and fittings (see p. 24).

2.6. TRANSFER LINES – PRINCIPLES

Figure 2.5 shows one end of a generalized transfer line. (These lines are sometimes called "siphons," but this term should be avoided, since it is

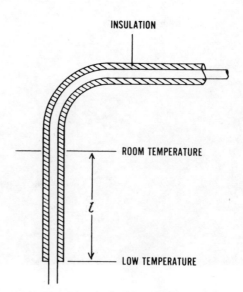

Fig. 2.5. Diagrammatic representation of part
of a cryogenic transfer line.

only exceptionally that siphonage in the strict sense of the word is physically possible.) The qualities to be taken into consideration are:

1. The heat influx to the inner tube must be appropriately small. The problems and their solutions are very similar to those discussed in Section 2.1; the maintenance of a low enough pressure is rarely a problem with liquid-hydrogen or liquid-helium lines, but thermal radiation is. There will also be a heat influx along the length l of the outside tube of the leg section which spans the gap between room temperature and that of the refrigerant. The same considerations therefore apply as in the case of the neck tube of a storage vessel.

2. Thermal contraction must not result in contact between the inner and outer tubes – notably at bends. This difficulty can be reduced by the use of alloys such as Nilo-36, but this material is expensive and not readily obtainable in the form of thin-walled tubing; various devices are generally used (see below) to accommodate contraction. Figures for

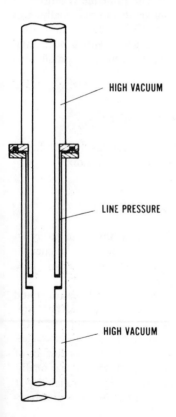

HIGH VACUUM

LINE PRESSURE

HIGH VACUUM

Fig. 2.6. The principle of the Johnston coupling.

these will be found in Chapter 7. A useful rule of thumb is that most common metals and alloys contract approximately 1 cm in 10 ft between room temperature and liquid-helium temperature.

3. Since it is rarely possible to precool a transfer line, the mass of the inner tube should be kept as low as possible in the case of liquid-helium lines. This point is illustrated by comparing two transfer lines of the sort commonly used to fill a cryostat from a storage vessel. A line in which the inner tube is made from thin-walled tubing usually takes less than 100 cm^3 of liquid helium to cool down. Certain proprietary lines are made flexible, and although this is an advantage in itself, some 2 litres of liquid helium may be required for cooling.

It is often necessary to incorporate a valve, for the reasons mentioned above. This can often be done by fitting an extenuated needle valve parallel with and attached to the outside of one leg.

Joints between transfer lines can be made by using the form of coupling described by Daunt and Johnston.[3] The principle is illustrated in Fig. 2.6; the only seal required is at room temperature. Ideally, this form of joint should be used in the orientation shown but if the mating parts are a good fit it is often possible to get away with a horizontal joint or even an inverted one. By quite simple means it is possible to combine a joint of this sort with a valve.

2.7. COMMERCIAL TRANSFER LINES

Many firms offer transfer lines complete or in sections for all liquid refrigerants, and these will be found listed in the Appendix.

2.8. LABORATORY-BUILT TRANSFER LINES

Transfer lines for liquid refrigerants can be made up quite easily in the laboratory with the advantages that they are then much cheaper than those available commercially, can be made to measure, and may even have a better performance.

There is a fundamental difference between transfer lines for liquid nitrogen, etc., and those for liquid hydrogen and helium. The latter cryopump themselves and, provided that there are no leaks from the inside tube into the insulating vacuum, no account need be taken of outgassing from the outer surface of the inner tube or from the inner surface of the outer. (The author has had cause to suspect that occasionally hydrogen is released from some materials, and for complete reliability it may be advisable to use the methods to be described for the construction of liquid-nitrogen lines for liquid-hydrogen lines as well, although usually those to be described for constructing liquid-helium lines will be found satisfactory.)

For liquid-nitrogen lines, where high efficiency is required or space is at a premium, the best type of insulation is high vacuum. However, as has been indicated, care has to be taken if a sufficiently low pressure is to be maintained indefinitely. The only technique which can be guaranteed to give satisfactory results is to pump out the line for not less than 24 hr while it is heated to about 500°C, e.g., by one of the commercially available heating tapes, before sealing off. A small quantity of charcoal may be included to adsorb any gas which despite this treatment might be released, but it is doubtful whether this is strictly necessary.

The necessity for this treatment restricts the possible means of construction: clearly, joints must be made by silver-soldering or welding and spacers cannot be made from plastics. Lines which have given good service for more than ten years have been made with 60/40 cupronickel tubing for the outside and copper tubing of the sort used for domestic plumbing (e.g., BSS 659) for the inside, with silver-soldered joints at the ends and for the pump-out branch. For bending they are loaded with coarse, dry sand. The final joint is made at one end, but a temporary plug is used at the other through which the inner tube can slide during bending. When the sand has been emptied out the final joint is made at the other end.

Powder insulation can be used to make lines having reasonably low thermal losses, but the thickness of insulation required – ~1 in – demands an outer tube which cannot be bent and may take up too much space. Solid or flexible foam insulants can also be used, but also suffer from the disadvantage of bulk (Chapter 7).

In the case of liquid-helium lines more possibilities offer themselves, including the use of soft solder for joints. However, there is also to be considered the necessity for low heat conductance along the outside tube of the end sections and the desirability of making the inner tube very thin-walled.

Some commonly used methods of fabrication are as follows:

1. Two pieces of thin-walled stainless steel ("cryogenic") tubing are arranged coaxially, loaded, and bent. A special bender has to be used because with this tubing a much larger radius is required for the bend than is usual for the diameter. A major disadvantage of such lines is their fragility even if stiffening webs are used at the bends. However, they have the advantage of being easy to make.

2. The outer tube can be made up out of a central section of relatively thick-walled copper or brass tube to include all parts which will be at room temperature when the transfer tube is complete, with

two thin-walled stainless steel ends. A thin-walled inner tube is then inserted and the whole loaded as before and bent at the extremities of the thick-walled portion.

3. A transfer line can be built up by using commercial pipe fittings for bends — these naturally have to be large enough for joints to be made to the inner tube, for which special elbows usually have to be machined. The result may be somewhat bulky, but this method of construction can be useful in some cases. (The transfer line in Fig. 2.4 has been made up in this way.)

4. Another method of construction (Croft and Jones[4]) involves attaching two short legs near to the ends of a length of copper tube. Thin-walled tubing is then hard-soldered to these legs and the complete inner tube inserted through a slot in the long copper tube, which is then closed by soft-soldering on a copper strip. End caps are then soft-soldered on.

A favourite method for sealing off transfer lines of all sorts is the copper/glass seal. Its advantages are that a seal of complete reliability can be made even by an amateur glassblower and that the thin section of the copper part of the seal provides a useful safety valve. A protective metal cap is required which must of course have a hole in it. Kovar/glass seals have the first advantage just mentioned, but not the second. Crimping of soft copper tubes with or without the use of internal solder also produces satisfactory seals, but success is more dependent on skill and experience.

The problem of allowing for contraction is greatest in the case of straight sections of transfer line. The only solution lies in incorporating a flexible bellows. The use of a bellows involves spacers which support the inner tube and which are also free to slide. However, with care these difficulties can be overcome. In the Clarendon Laboratory we have recently constructed a 40-ft ½-in. i.d. liquid-hydrogen line (Croft[5]) composed largely of copper tubing. Contraction is taken up by a 7-in. section of brass bellows prevented from snaking by a loosely-fitting PTFE tube surrounding them. Low-conducting spacers at 5-ft intervals have PTFE bosses at their room-temperature ends so that they are free to slide in the 2-in. outer tube.

However, it is more usual for transfer lines to be shorter and to include at least one right angle — usually two. There are various tricks which can then be used:

1. In a fabricated transfer line the inner tubes can be made longer than is consistent with their being coaxial with the outer tubes at room temperature.

2. In addition, bushes at the ends of the legs can be made eccentric.

3. Where a pair of tubes start by being straight and coaxial the

inner tube can subsequently be pushed up before joints are made at the ends so as to give more space on the corners.

4. In addition, a sleeve joint can be made at the centre such that the ends of the tubes are a few millimetres apart. After bending, this joint can then be reheated and the ends butted up so as further to increase the clearance on the corners.

As mentioned above, a well-designed and carefully made liquid-helium transfer line should not take more than 100 cm^3 of liquid helium to cool down. Sometimes transfer lines are found to need much more than this or even to refuse to transmit liquid helium at all. The trouble is nearly always an internal touch and this can be checked by an X-ray photograph. Fabricated transfer lines which do not need spacers can be checked by tapping them: a characteristic tinkle proves that the inside tube is free, although, of course, it may not remain free at low temperatures. Another cause of trouble with liquid-helium transfer lines is too short a length of thin-walled stainless steel tubing for the double-walled portion of the leg – 12 in. seems to be the minimum satisfactory length. Finally, a case has been reported of failure of a transfer line to transmit which was finally traced by means of an X-ray photograph to the flattening of a short length of the inner tube caused by an unnoticed temporary overpressure in the vacuum space during fabrication.

2.9. FLOW PROMOTION

Where a liquid refrigerant is drawn from a storage vessel the latter should be maintained at the lowest pressure consistent with an adequate flow rate. It is in the case of liquid helium that it is the most important to keep this overpressure as low as possible since, on account of its small latent heat compared with its specific heat, a relatively high proportion will evaporate – known as the "flash loss" – on reduction of pressure by a given amount. The overpressure used should not exceed 0.1 atm. In the case of liquid nitrogen overpressures of up to 1 atm are usual. Overpressures sometimes maintain themselves by the evaporation due to the normal heat influx to the vessel. If this is insufficient, heat can be introduced from outside – e.g., by means of the pressure-raising coil mentioned above in the case of large liquid-nitrogen vessels, by artificially increasing the heat thermally conducted into smaller vessels containing any refrigerant, or by an electrical heater. A useful means of introducing small quantities of heat into a liquid-helium vessel is by applying pulsating pressure to a rubber football bladder connected to it near to the point where the transfer line leaves the vessel.

In rare cases pumps are required for liquid nitrogen, and manufacturers of these will be found listed in the Appendix. (See also Chapter 6, p.116 for pumps incorporated in automatic refrigerant control systems.)

Valves for the control of liquid refrigerants are described in Chapter 7.

REFERENCES

1. Wexler, A., *J. Appl. Phys.* **22**, 1463 (1951); Wexler, A., and Jacket, H. S., *Rev. Sci. Instr.* **22**, 282 (1951).
2. White, G. K., *Experimental Techniques in Low-Temperature Physics*, 2nd ed., Oxford University Press (Clarendon Press) (1968), p.211.
3. Daunt, J. C., and Johnston, H. L., *Rev. Sci. Instr.* **20**, 122 (1949).
4. Croft, A. J., and Jones, G. O., *Brit. J. Appl. Phys.*, **1**, 137 (1950).
5. Croft, A. J., accepted for publication in *Cryogenics* (1970).

Note added in proof:

Thermo-acoustic oscillations of the type described on p.110 sometimes occur in storage vessels and cryostat dewars, and can give rise to abnormally high evaporation rates. They can usually be cured readily by introducing acoustically absorbent material.

Chapter 3

Liquefiers and Refrigerators

3.1. PRINCIPLES

There are three ways of cooling a gas: (1) by exploiting the *latent heat of evaporation* of another gas of lower boiling point; (2) by causing it to do *external work*; and (3) by making use of the negative Joule — Thomson effect, arising from repulsive intermolecular forces which result in the performance of what is called *internal work* on expansion.

The first of these is the basis of the *cascade* principle, which depends upon a succession of evaporative refrigeration systems in which each working substance has a critical temperature higher than the triple-point temperature – or more usually the boiling point at a pressure just above atmospheric – of that of the next warmer substance, and a triple-point temperature lower than the critical temperature of the next colder. An example of such a system is that designed by Keesom* for the liquefaction of nitrogen: ammonia–ethylene–methane–nitrogen. Such a system can have a high thermodynamic efficiency, but is too cumbersome for laboratory use. Moreover, this principle is not applicable to the liquefaction of neon, hydrogen, and helium because of the wide temperature gaps between the triple-point temperature of oxygen and the critical temperatures of neon and hydrogen, and between the triple-point temperatures of these two substances and the critical temperature of helium (see Table I). The cascade principle is therefore of no help in reaching very low temperatures. (The term "cascade" should not be applied to liquefaction systems in which, for example, a helium liquefier embodies a hydrogen refrigeration stage: it should be applied only to systems depending on change of state.)

The cooling which occurs when a gas expands and performs external work has been applied to cryogenics in three distinct ways:

1. A work-absorbing device is included in a *continuous* closed

---

* See, for example, Davies.[1]

circuit – usually of helium gas. This device may be a piston and cylinder or a turbine.

2. Similar in principle are systems which work on a cycle which is usually identical with or similar to the classical Kirk cycle, which is the reversed refrigeration equivalent of the earlier Stirling heat-engine cycle (see Collins and Cannaday[2]).

3. In the Simon liquefier, which is in a class of its own since it operates *discontinuously*, a batch of liquid helium is produced in the course of one circuit of a refrigeration cycle.

In contrast with the Joule–Thomson effect, external-work cooling occurs with all gases under all conditions of temperature and pressure.

The Joule–Thomson effect is the change in temperature which occurs when the pressure of a gas changes but no work is done; in practice, a fall in pressure usually occurs at a valve. The change in temperature may be either positive or negative, depending on whether the forces between the molecules tend to bring them together or keep them apart. For every substance there is an *inversion* temperature at which the forces between molecules are zero and no Joule–Thomson effect occurs. Below this temperature cooling occurs on expansion and the phenomenon can then be used as a means of cooling the gas. Table IX gives the inversion temperature of some common gases. The temperature to which a gas must be *precooled* if a reasonable proportion of it is to liquefy is about one third of the inversion temperature. It will be seen that of the five gases in Table IX, only oxygen and nitrogen can be liquefied from room temperature. Nearly all liquefiers embody a final stage of Joule–Thomson cooling; the difference between the various types of hydrogen and helium liquefier lies in the means used for precooling. We shall first consider this final stage – taking the case of helium – and then review the possibilities for the rest of a liquefaction or refrigeration system.

TABLE IX

**Joule–Thomson Effect
Inversion Temperatures**

Gas	T_{inv} ($^{\circ}$K)
Oxygen	893
Nitrogen	621
Neon	260
Hydrogen	205
Helium	51

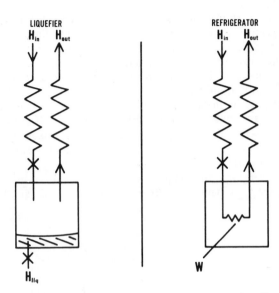

Fig. 3.1. Final stage of internal-work (Linde) liquefier
and refrigerator.

All that is essential to a liquefier and a refrigerator appears in Fig.
3.1. The pairs of zig-zag lines represent heat exchangers: devices in
which heat can pass from a stream of a gas at a higher temperature to a
different stream at a lower temperature. It can be taken that it is
realistic to assume equal temperatures for ingoing and outgoing gas
streams when calculating performances of liquefiers and refrigerators.
The cross represents a needle valve, across which there is a change of
pressure – typically, from anything between 20 and 120 atm to approximately 1 atm according to the working substance. Taking first the case
of a liquefier, let us define x as the fraction of the mass flow rate of
helium gas which is liquefied and assume that this liquid is continuously
withdrawn from the system as it is made. (When comparing the
performance of actual liquefiers with that predicted from thermal data
one must – especially in the case of helium – distinguish this condition
from the usual state of affairs, i.e., the filling up of a reservoir with the
consequent displacement of cold vapour.) Taking the case of unit mass
flow rate we can equate the enthalpies entering and leaving the system
as follows:

$$H_{in} = xH_{liq} + (1-x)H_{out}$$

or

$$x = (H_{out} - H_{in})/ (H_{out} - H_{liq})$$

The full curves shown in Figs. 3.2 and 3.3 give for hydrogen and helium, respectively, the values of x plotted against the pressure of the incoming gas stream for various values of its temperature on the assumption that the pressure above the liquid is 1 atm. It will be noticed that there is a somewhat flat peak in the curves and that the optimum pressure falls with temperature.

In the case of the refrigerator shown on the right in Fig. 3.1 the power which can be absorbed is given by $W = H_{out} - H_{in}$ watts per unit mass flow rate, where the enthalpies are in J-g^{-1}. The broken curves in Figs. 3.2 and 3.3 give, by reference to the right-hand vertical axis, the refrigeration in watts for unit mass flow.

It is interesting to compare the liquid yield in the two cases. Taking 1 g-sec^{-1} of helium entering a liquefier circuit at 30 atm and

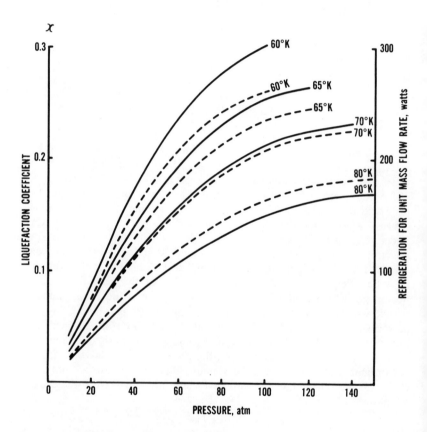

Fig. 3.2. Liquefaction coefficient and refrigerative power for unit mass flow for hydrogen. Full curves: liquefaction coefficient. Broken curves: refrigerative power.

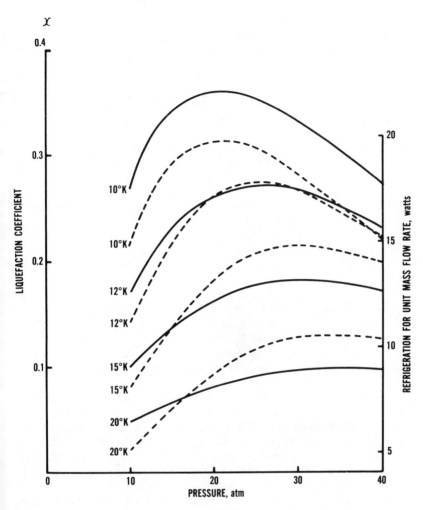

Fig. 3.3. Liquefaction coefficient and refrigerative power for unit mass flow for helium. Full curves: liquefaction coefficient. Broken curves: refrigerative power.

15°K we find a liquefaction rate of 5.2 litres-hr^{-1} (1 g-sec^{-1} corresponds to a flow rate at NTP of 20.2 m^3- hr^{-1} which in practice would call for a compressor absorbing roughly 7 h.p.) In the case of a refrigerator taking the same flow under the same conditions of temperature and pressure, we find that 14.7 W can be absorbed, corresponding to an evolution of liquid of 20 litres-hr^{-1}.

The rest of a liquefier or refrigerator system is concerned with: (1) cooling the working substance at an appropriate pressure to a temperature at which use can be made of Joule–Thomson cooling, (2) the recovery of as much as possible of the cold of the working substance when it leaves the final heat exchanger after expansion to approximately 1 atm, (3) purification of the gas stream so that solid impurities will not be deposited, and (4) in the case of hydrogen systems the promotion of conversion of *ortho*-hydrogen should this be required (see p. 6).

A hydrogen system can be precooled by means of a bath of liquid nitrogen. This can be pumped down to any temperature above the triple point, and it will be seen from Fig. 3.2 that since the performance improves rapidly with decrease in precooling temperature, it is advantageous to get as near to the triple point as possible. Such an arrangement provides a simple design for a hydrogen liquefier or refrigerator on a laboratory scale, but the inherent inefficiency from the point of view of power consumption rules it out for large-scale plant.

In the case of helium systems, liquid hydrogen under reduced pressure similarly provides a satisfactory means of precooling. Neon is unattractive as a precooling agent, since at its triple point of $24.6°K$ the value of x for helium from 35 atm is only 0.053 — i.e., a flow rate of helium four times greater would be required for a liquefier precooled by liquid neon than for one precooled by liquid hydrogen.

The alternative is to cause the working substance to cool by making it do external work. In many cases it is obvious where the external work is done — e.g., on the piston of an expansion engine or on the blades of a turbine. In some systems it is less clear, and one has to look carefully to discover where the work is being performed—e.g., the Cryodyne cycle below. The precooling temperature reached in a helium expansion engine system can be considerably lower than $14°K$ — the triple point of hydrogen, and therefore the minimum temperature at which cooling by means of liquid hydrogen is possible. In the ADL–Collins helium liquefier (see below) helium enters the Joule–Thomson stage at $10°K$. For this temperature and the optimum pressure of 19 atm the value of x is 0.42.

Turning to the question of the conservation of cold, we see from Fig. 3.1 that in a liquefier we have a fraction $1 - x$ of the gas passing through the system, and in a refrigerator the whole of it, leaving the final stage at close to the precooling temperature and at approximately 1 atm pressure. It is clearly essential that as much as possible of this cold should be recovered. The obvious course is to use a heat exchanger to cool down that quantity of the incoming warm gas the change in enthalpy of which is equal to the change in enthalpy of the returning cold gas between the temperature at which it leaves the final heat exchanger and room temperature. If m_{in} and m_{out} are the mass flow rates in such a heat exchanger, then this so-called balanced condition obtains when

$$m_{in} \, \Delta H_{in} = m_{out} \, \Delta H_{out}$$

where ΔH_{in} and ΔH_{out} are the changes in enthalpy from room temperature to the precooling temperature for the high- and low-pressure gas streams, respectively. The straightforward application of these principles can give rise to misleading results in the case of hydrogen, in which there is a change from the specific heat of a diatomic gas $(^7/_2 R)$ to that of a monatomic gas $(^5/_2 R)$ over the temperature range $80-180°K$. Such a case arises if hydrogen evaporated from a precooling bath at, say, $15°K$ is used to cool a stream of helium gas starting from room temperature. This difficulty has been discussed by Mann *et al.*[3]

In the case of a liquefier it will be found that a fraction *approximately* equal to $(1 - x)$ of the total mass flow can be passed down this heat exchanger — not exactly equal to $(1 - x)$ because of the difference in enthalpies between the gas at high and low pressures, which accounts for the cooling on expansion. The fraction approximately equal to x has to be cooled entirely by the precooling arrangements — whatever they may be. In a refrigerator system the fraction of the total flow which can be cooled by the returning gas at approximately atmospheric pressure is of course very much larger, as the following enthalpy values show:

Helium gas at 30 atm. Change in enthalpy from $280°K$ to $15°K$: 1402 $J\text{-}g^{-1}$.

Helium gas at 1 atm. Change in enthalpy over the same temperature range: 1377 $J\text{-}g^{-1}$.

From these figures we see that a fraction 0.98 of the incoming high-pressure gas can be cooled by the returning gas stream if we assume a perfect heat exchanger.

In a system where a gas stream is cooled by means of, say, liquid nitrogen a heat exchanger can be used to conserve the cold of the nitrogen vapour. In this specific case the inclusion of such an exchanger will reduce the consumption of liquid nitrogen by a factor of ~2.

Purification in general is considered in Chapter 5. Liquefiers and refrigerators usually embody vessels containing such adsorbents as activated charcoal to remove residual impurities which cannot conveniently be removed at room temperature. As a rule, such purifiers function at the normal boiling point of the substance which they are required to remove, but performance improves rapidly with fall in temperature. In a hydrogen system, e.g., it is as well to have a smaller, secondary purifying stage working at the precooling temperature — e.g., $65°K$ — as well as the main purifier working at $77°K$. In a helium system where most of the helium is recovered it is important to have a purifier working at $20°K$ or less to take out the residual neon impurity usually present in mineral helium gas; otherwise there may be a considerable build up of neon in the system over a period of time. Although it is very unlikely that a quantity of helium could get into a hydrogen system, the consequent effect on its performance would be so open to wrong diagnosis that it seems worth mentioning the possibility.

As we have seen in Chapter 1 (p. 7), it is unlikely that a laboratory based hydrogen liquefier will be required to produce liquid *para*-hydrogen, and conversion of *ortho*-hydrogen is undesirable in a refrigerator, since it merely increases the running cost. We shall therefore not go into the question of *ortho-para* catalysis (see Scott *et al.*[4]).

3.2. COMMERCIAL EXTERNAL-WORK LIQUEFIERS AND REFRIGERATORS

The ADL–Collins Helium Liquefiers and Refrigerators

The appearance on the market in 1946 of a package helium liquefier developed by Arthur D. Little, Inc.[Ar]* from a design by Prof. Samuel Collins of MIT marked the beginning of a rapid spread of low-temperature work which is still continuing. More than 300 have been built, and although they are less of a keystone in the cryogenic world than they were, owing to the increasing use of purchased liquid helium, the Collins liquefiers still maintain their place in spite of increasing competition. An account of the development of the expansion engine used will be found in the work by Collins and Cannaday[2] and in a paper by Collins.[5] There have been various modifications in detail since the liquefier was first marketed, but it remains the same in essence.

Two piston and cylinder expanders are used (see Fig. 3.4) working between $60°$ and $30°K$ and between $15°$ and $9°K$ (these are approximate figures). The incoming helium gas is distributed roughly as follows (Fig. 3.5):

First engine	30%
Second engine	55%
Final Joule–Thomson stage	15%

The engines themselves are of ingenious design. The piston rod is always in tension and can therefore be light and flexible. This has two important results: it conducts a negligible amount of heat from the room-temperature crankshaft, and also it allows the piston to take up a coaxial position in the cylinder. Piston and cylinder are made of nitrided nitralloy steel† and the clearance is about 0.002 in. on a bore of 1 in.

---

*See Appendix for address.
†See footnote to p. 37.

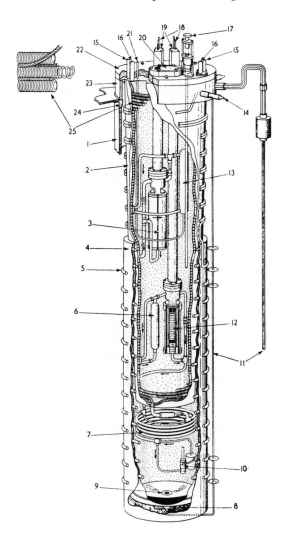

Fig. 3.4. Interior of ADL–Collins helium liquefier. Key: 1: Outer vacuum shell. 2: Inner vacuum shell. 3: No. 1 expansion engine. 4: Radiation shield. 5: Auxiliary precooling coil. 6: No. 2 charcoal pot. 7: Joule–Thomson heat exchanger. 8: Bulb of helium thermometer. 9: Partition plug. 10: Expansion valve body. 11: Liquid drawing-off tube. 12: No. 2 expansion engine. 13: No. 1 charcoal pot. 14: Auxiliary liquefier inlet. 15: Compressed helium inlet. 16: Expanded helium outlet. 17: Expansion valve knob. 18: Piston and valve rods. 19: Stuffing boxes. 20: Experimental chamber opening. 21: No. 1 charcoal pot blow-down. 22: No. 3 flange. 23: No. 2 flange. 24: No. 1 flange. 25: Main heat exchanger. (Drawing: Arthur D. Little, Inc.)

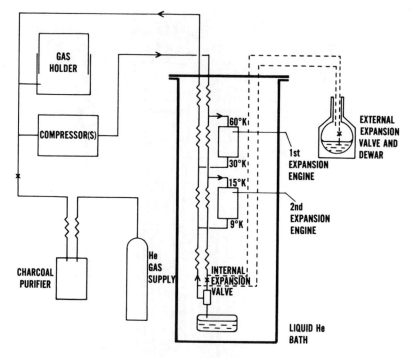

Fig. 3.5. Flow diagram of ADL–Collins helium liquefier.

Peripheral grooves in the piston help it to assume a stable position. The inlet and exhaust valves have leather seats and their operating rods are also in tension. At the top of the machine are the stuffing boxes, crossheads, crankshafts, and flywheels – an independent set for each engine.* Power is absorbed hydraulically.

The heat exchanger is of the type in which a helix of finned tube is enclosed between two near-cylindrical, thin, stainless steel sheets (in fact, these are frustums of cones of small angle). The outer shell of the main heat exchanger is extended downwards so as to form an approximate cylinder. At the bottom of this is the reservoir of liquid helium, and above it – in a stagnant atmosphere of helium gas – are the expansion valve, the final heat exchanger, and the two expansion engines. Surrounding the bottom half of the main heat exchanger and the enclosure below it is a radiation shield which is in thermal contact with the heat exchanger, but provision is made for cooling it with liquid nitrogen. In this case the helium gas stream is cooled additionally by the liquid nitrogen, and, as will be seen below, the output of the liquefier is

* See also footnote to p. 37.

thereby doubled. An outer cylindrical tank enables a high vacuum to be maintained in the space surrounding the enclosure described above and including the radiation shield. This construction is equivalent to putting the whole of the helium system into a dewar vessel and has the advantage that there are few sources of possible leaks into the high-vacuum insulation.

The standard Kellog compressor unit supplies 35 ft^3-min^{-1} at 220 lb-in^{-2} and consists of two separate units, each of two stages, which together form an air-cooled four-stage compressor. These two units are driven by a common centrally-placed driving motor. (It is making heavy weather of compressing helium to 15 atm to use four stages. At the Clarendon Laboratory we have for many years been using two-stage compressors for helium to 27 atm and hydrogen to 66 atm — see Chapter 4, p.80 — with no trouble attributable to oil decomposition. These four-stage compressors therefore seem unnecessarily bulky and expensive.)

The ADL–Collins liquefier is available in the following forms:

Model 25. Includes one 35 ft^3-min^{-1} compressor set and liquefies into a 15-litre internal reservoir from which liquid is withdrawn discontinuously by hand. Output: Without liquid nitrogen: 2 litres-hr^{-1}. With liquid nitrogen: 5 litres-hr^{-1}.

Model 50. As above, but with two compressor sets. Output: Without liquid nitrogen: 3 litres-hr^{-1}. With liquid nitrogen: 9 litres-hr^{-1}.

Model 100. Includes one compressor set and has the same output as Model 25, but, by means of a double transfer line with a secondary expansion valve at the end remote from the liquefier, it is possible to liquefy continuously into an external storage vessel and thus to run the liquefier unattended for long periods.

Model 200. Includes two compressor sets and has the same output as Model 50, but is fitted with the same arrangement for liquefying into an external storage vessel as Model 100.*

Requirements	Power (kW)	Cooling water (gal-min^{-1})	Liquid N_2 where used (litres-W^{-1})
Models 25 and 100	13	5	10–12
Models 50 and 200	26	10	10–12

---

* *Recent development.* These four models now incorporate an expansion engine made from plastic materials and having a greater piston-to-cylinder clearance. The crosshead is electromechanical and is quieter in operation. These new expanders and crossheads are available for fitting in to existing liquefiers. The new liquefiers are called Models 1025, 1050, 1100, and 1200 and are in other respects similar to those described above.

Fig. 3.6. Model 200 ADL–Collins helium liquefier. (Photo: Arthur D. Little, Inc.)

Figure 3.6 shows the Model 200 with the 250-litre storage vessel. The two compressor sets and the dry-seal gasholder can also be seen. These liquefiers are fitted with the necessary equipment to shut them down should abnormal conditions arise while they are running unattended. A purifying unit for the make-up helium gas is supplied. (For helium recovery ADL manufacture a separate unit which is described in Chapter 5.)

Modifications of this basic design provide for the condensation of hydrogen (see Scott et al.[4] for details) and for use as a refrigerator rather than as a liquefier.

Enquiries about the reliability of these machines from a somewhat limited number of users suggest the following conclusions: the liquefier itself has high reliability provided it is looked after with unremitting care. Considerable damage can result from excessive impurities in the gas stream, and skilled attention is required to such matters as the degree of compression on the stuffing boxes through which the piston rods and valve-operating rods pass. Vacuum leaks seem to be rare. Reports are not uncommon of mechanical failure in the compressors despite careful treatment, but improvements are being incorporated in current equipment. No better commendation of these liquefiers can be offered than the fact that the two firms which supply all the liquid helium sold in Great Britain at the time of writing use two of these liquefiers each, as does AERE Harwell.

Arthur D. Little, Inc. in cooperation with Prof. Sam Collins have recently introduced a helium liquefier producing 1 litre-hr^{-1} — the Model 520. Apart from being a miniaturized version of the original ADL— Collins liquefier, this has certain basic differences in design which in the writer's opinion are improvements. The compressor is a specially-designed two-stage sealed unit delivering 14 ft^3-min^{-1} at 250 lb-in^{-2}. This is far more compact than the compressor sets supplied with the helium liquefiers just described and, moreover, has half as many pistons, valves, big-end bearings, etc. The original Collins liquefier had two expansion engines so that it could be run without the use of liquid nitrogen, but it must very rarely be the case that one is so run. In this new liquefier there is only one expander and the use of 4-6 litres-hr^{-1} of liquid nitrogen is unavoidable. The internal liquid-helium reservoir holds 15 litres. Withdrawal has to be done by hand, but the liquefier is suitably protected for unattended running. Its requirements are 12 kW of electrical power, 4 gal-min^{-1} of cooling water. As Fig. 3.7 shows, the whole system, apart from the 9 ft^3 dry-sealed gasholder, is in one unit. It measures 78 in. by 32 in. and weighs 1950 lb, and requires 11 ft ceiling clearance for maintenance. There is every reason to suppose that this liquefier would be ideally suitable where the demand for liquid helium is limited to about 15 litres/day.

Fig. 3.7. Model 520 ADL–Collins helium liquefier. (Photo: Arthur D. Little, Inc.)

The Linde Company's Helium Liquefiers and Refrigerators

The Linde Co.[Li] of West Germany — not to be confused with the Linde Division of Union Carbide, Inc. — have developed a helium liquefier which is similar in principle to the mini-Collins just described in that it embodies one stage of external work in one piston and cylinder expander using liquid nitrogen for the initial cooling. The expansion engine has been developed by Doll and Eder[6] (see Fig. 3.8). The place of inlet and exhaust valves is taken by ports in the cylinder wall in somewhat the same way as in a two-stroke petrol engine. When the piston reaches the bottom of its stroke an annular groove A near the top of the piston meets a similar groove in the cylinder wall B which is fed with high-pressure helium gas at about $26°K$ through a number of small holes. This gas passes through radial holes C in the piston D into the axial space E inside it. As the expansion stroke begins the entry of gas is cut off and the piston continues to rise until its lower extremity clears the annular groove F, when the expanded gas in the cylinder leaves, at about 14°K. In this expansion engine, in contrast with the Collins, the piston rod G is in compression, not in tension — freedom of the piston to take up a stable position in the cylinder H is achieved by providing a convex surface on the axis of the piston which rolls upon a flat surface at the lower end of the piston rod, being maintained in contact with it by means of a spring. Gas lubrication and stabilization of the piston are achieved by a complex treatment of the piston surface. (A full description will be found in another paper by Doll and Eder.[7])

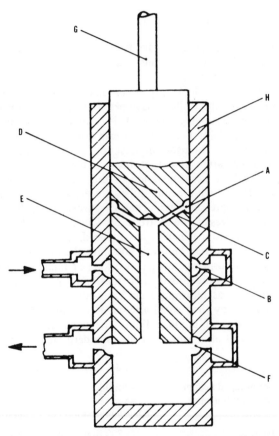

Fig. 3.8. The principle of the Doll–Eder expansion engine. Key: A – Inlet peripheral groove in piston. B – Inlet port in cylinder wall. C – Radial holes in piston. D – Piston. E – Axial space inside piston. F – Exhaust port in cylinder wall. G – Piston rod. H – Cylinder wall.

Unless the clearance between the piston and the cylinder is very small there will be a continuous short-circuiting flow of gas from the high-pressure to the low-pressure port. The following figures show the effect of the clearance between the piston and the cylinder on the leakage rate of the gas when the engine is at its operating temperature.

Clearance (μ)	Inlet pressure (atm)	Leakage rate ($m^3 \cdot hr^{-1} \cdot NTP$)
3.5	20	13.6
1.7	26	3.8
1.3	27	1.4

The clearances used in practice vary from 1.0 to 2.0 μ (compare this with the clearance in the current models of the larger ADL–Collins expansion engines, which is about 50 μ).

The circuit of these liquefiers and refrigerators is similar in principle to that of the Collins machines. Helium at 20 atm from one or two (see below) four-stage air-cooled compressors is split into two approximately equal fractions: one stream goes to the expander and the other to the Joule–Thomson liquefaction circuit, both having passed through the usual arrangement of heat exchangers and liquid-nitrogen precooling. The heat exchangers are of the classical pattern in which a number of parallel high-pressure tubes are enclosed by a tube carrying the returning low-pressure gas. Thermal insulation is by means of a cylindrical dewar vessel the bottom of which forms the reservoir for liquid helium and from which it has to be drawn off about every 2 hr. The helium gas is purified before it goes into the liquefier in one of two interchangeable adsorptive purifiers working at liquid-nitrogen temperature. Two gasholders are provided so that one can be used for clean gas and the other for gas returning from cryostats, etc., but their usefulness is severely limited by their small size, namely 1 m^3, which is equivalent to about 1.4 litres of liquid helium. The compressors are capable of working up to 130 atm, which enables standard storage cylinders to be

Specification	Model VR4	Model VR8*
Output of liquid helium	4 litres-hr^{-1}	8 litres-hr^{-1}
Helium gas throughput	45 m^3-hr^{-1}	90 m^3-hr^{-1}
	(One compressor)	(Two compressors)
Engine speed	7–800 rpm	11–1200 rpm
Liquid nitrogen consumption	5–6 litres-hr^{-1}	10 litres-hr^{-1}
Electrical power consumption	9 kW	18 kW
Start-up time	approx. 50 min.	100 min.
Space required	20 m^2	24 m^2

*Shown in Fig. 3.9.

used to their full pressure. The instrumentation is liberal and includes a recorder showing the purity of the helium gas. However, these liquefiers cannot be left running unattended for long periods.

The model V20 produces 20 litres-hr^{-1} and embodies two expansion engines working in tandem. Two separate compressors are used for the expansion engine and liquefaction circuits, the latter being capable of working up to 130 atm. The cool-down time is 150 min. These three liquefiers can be used as refrigerators to absorb 15, 30, and 75 W, respectively, at 4.4°K.

Eighteen of these liquefiers have been supplied to European laboratories. Recently the piston has been modified so as to include polytetrafluorethylene rings: these must reduce the possibility of seizure, but it is not clear how the very small leakage rate past the piston is maintained. Reports from users suggest that these liquefiers work well provided that the helium gas is very pure: quite small levels of impurity cause the engine to seize up, in which case the makers have to

Fig. 3.9. Model VR8 helium liquefier manufactured by Linde's Eismaschinen Aktiengesellschaft. (Photo: Linde Aktienges.)

fit a replacement. The purifying arrangements are regarded as being of insufficient capacity for long runs.

The Philips Liquefiers and Refrigerators

This important range of equipment is based on a device which, together with other equipment discussed later, is fundamentally an application of the heat-engine cycle invented by Stirling in 1827. The Stirling cycle was first used in reverse as a refrigeration cycle by Kirk in *c*. 1861. It is shown on a pV diagram in Fig. 3.10a and in diagrammatic form in Fig. 3.10b. Starting at *a* the working substance is compressed isothermally at room temperature to *b*, with the evolution of heat. It then passes through a component called a regenerator; in reality, it is a container filled with solid packing material to which the working substance can transfer heat or *vice versa*. For the purposes of this theoretical treatment it must be considered to have zero dead volume for gas. In this case we assume that the regenerator has been left below the temperature of the working substance at *b* in the course of the preceding cycle. The working substance is cooled in this regenerator at constant volume to point *c* on the cycle. Isothermal expansion then takes place in the other piston and cylinder at low temperature until the gas returns to its original volume at *d*. In doing so the working substance absorbs heat. The gas is then transferred through the same regenerator and again at constant volume to its original state at *a*. This cycle is the basis of several cryogenic devices, although, not unnaturally, none follow it rigorously. For engineering reasons it is sometimes necessary for pistons to execute harmonic motions rather than the discontinuous motions of the idealized Kirk refrigeration cycle just described. Further, it is often the case that compression and expansion come closer to being adiabatic (isentropic) rather than isothermal – naturally, the more so the shorter the cycling time. However, the principal difficulty in applying thermodynamics to real systems arises from the finite volume of the regenerator.

A regenerator performs a similar function in a cyclic process to that of a countercurrent heat exchanger in a continuous process. An optimal regenerator has the following properties:

1. Minimum resistance to fluid flow.
2. Minimum longitudinal thermal conductance.
3. Maximum thermal conductivity of the packing material.
4. Maximum heat transfer rate between the fluid and the packing material.
5. Maximum specific heat of the same.

A

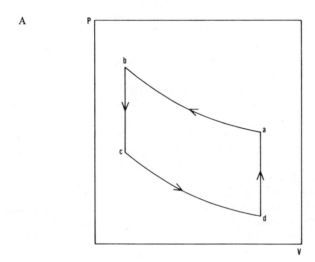

B

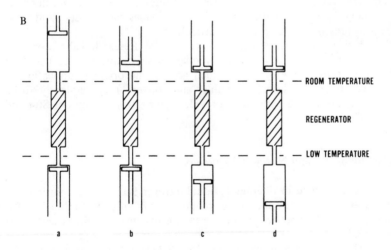

Fig. 3.10. Principle of Kirk or reverse Stirling refrigeration cycle. (A) pV diagram. (B) Positions of moving parts at beginning of each stage.

In practice, resistance to fluid flow can be made negligible by choosing an appropriate cross-sectional area for the column. Thermal contact between the pieces of packing material will usually be sufficiently bad for the longitudinal thermal conductance to be sufficiently low. High thermal conductivity of the packing material is of little importance if the material is in a finely-divided state, and this also aids heat transfer between it and the fluid. The question of the heat capacity of the packing material calls for closer examination. The heat capacity per unit volume is very roughly the same for most substances at temperatures not greatly below room temperature – the exceptions being of course substances with a high Debye θ, which have a specific heat less than the classical value at room temperature or at temperatures well above say $100°K$. Many substances will therefore do for this temperature range – indeed, a famous Swiss firm are said to pack their regenerators with the cherry stones discarded by jam manufacturers. At lower temperatures one must look carefully at the heat capacity temperature curves when choosing regenerator packing. Copper, for example, has lost only half of its classical high-temperature heat capacity at about $80°K$, and is therefore a satisfactory packing material for a regenerator working down to about the temperature of liquid nitrogen, but not very much lower. In the case of lead the corresponding temperature is about $25°K$, and it is therefore useful for regenerators working down to liquid-hydrogen temperatures. Still lower temperatures present problems in that only substances with specific heat anomalies can be used, and these tend to present serious practical difficulties.

Köhler and his associates at the Philips Research Laboratories at Eindhoven[Ph] have developed a number of cryogenic devices based on an adaptation of the cycle described above. The original application of this was to an air liquefier (Köhler and Jonkers[8]), subsequently fitted with a fractionating column to provide liquid nitrogen (van der Ster and Köhler[9]).

The Philips Air and Nitrogen Liquefiers

As explained in Chapter 1, liquid nitrogen is usually bought in bulk. However, where convenience or economics or both favour the installation of a local liquefier one of the Philips machines is almost certain to be found.

The expansion machine (see Fig. 3.11) embodies two pistonlike components which are caused to move up and down by means of a crankshaft. It is important to distinguish the functions of these two components: the lower annular piston is concerned with performing and absorbing work, while the upper one is concerned solely with moving

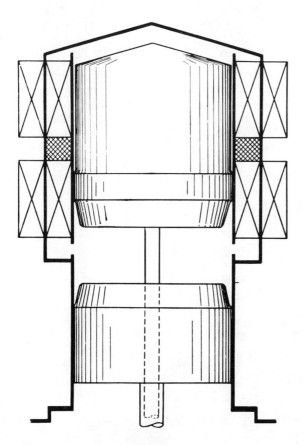

Fig. 3.11. Single-stage Philips refrigeration machine. (Drawing: Philips, Eindhoven.)

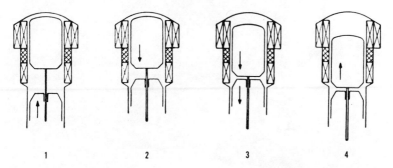

Fig. 3.12. Stages of Philips refrigeration cycle. (Drawing: Philips, Eindhoven.)

gas at almost constant pressure from one part of the system to another — doing or absorbing negligible work — and is therefore called the displacer. The operating cycle is as follows (see Fig. 3.12):

1. *Compression.* The working substance — hydrogen or helium gas — is compressed by the piston into the compression chamber. The heat of compression is absorbed by cooling water.

2. *Transfer from hot part of system to cold part.* The displacer then falls so that the gas passes through the regenerator, which is packed with fine phosphor-bronze wire. Here it picks up any cold dumped during the previous cycle and passes to the expansion chamber at the top.

3. *Expansion.* The piston then falls, and, on expanding, the gas does work on it and therefore cools. At the same time the displacer moves further downwards with the piston.

Fig. 3.13. Model PLN 430 Philips nitrogen liquefier. (Photo: Philips, Eindhoven.)

4. *Transfer from cold part of system to hot part.* The displacer then rises to its fullest extent, causing the gas to return to the compression chamber through the regenerator, where it leaves what remains of its cold.

The top of the expansion chamber is therefore the heat sink in this system, as will be seen from Fig. 3.11, and this is where air condensation surfaces are fitted. The displacer has a low longitudinal thermal conductivity, and only the lower part is in contact with the cylinder wall. The engine is directly coupled to an electric motor running at 1450 rpm. Four expanders have been ganged together to form one unit, and this is shown in Fig. 3.13. The specifications of the single- and four-expander air and nitrogen liquefiers are given as follows:

Output

Model	
PLA 107:	liquid air-7.5 litres-hr^{-1}
PLN 106:	liquid nitrogen-6 litres-hr^{-1}
PLA 433:	liquid air-33 litres-hr^{-1}
PLN 430:	liquid nitrogen-30 litres-hr^{-1}

Requirements

Model	Power (kW)	Cooling water (m^3-hr^{-1})
PLA 107, PLN 106	9	0.75 (gal-min^{-1})
PLA 433, PLN 430	36	3.00 (gal-min^{-1})

Refrigerators based on the single- and four-expander machines are available:

Model PGA 105   power absorbed: 1 kW at $-180°$C
Model "B"   power absorbed: 4 kW at $-180°$C.

The input power and cooling water requirements are the same as for the corresponding liquefiers.

The start-up times for the four liquefiers range from 30 to 90 min. The time for which they can be run before defrosting is required naturally varies with atmospheric humidity, but can be as long as six days. The time taken to defrost varies from 2 to 6 hr.

As in the case of the ADL—Collins helium liquefier, these machines give excellent service if they are carefully looked after. An Oxford laboratory has had one in more or less continuous use for over ten years. It is completely stripped down and reassembled by a member of the laboratory staff once a year and spare parts fitted where there are any signs of wear. This is not to say that the service provided by the makers is not excellent, but rather to indicate that the machine is such that it can be maintained indefinitely by a sufficiently capable technician.

Liquefiers and refrigerators of similar specifications are made in the USA by North American Philips Co., Inc.(No)

The Philips Two-Stage Cryogenerator Model PEH-100

Prast,[10] of the Philips Research Laboratories, has described a development of the reverse Stirling cycle expansion engine which enables much lower temperatures to be reached. The expansion chamber is split into two parts with a second displacer mounted above a first displacer which is similar to that in the original machine. There is also a second regenerator between the top or secondary expansion space and the intermediate expansion space – this contains fine lead shot. Refrigeration loads may be applied to both the low- and intermediate-temperature heat sinks, and representative figures for the loads which can be sustained are given as follows:

	Temperature of second-stage heat sink ($^\circ$K)				
	15	20	30	40	50
Power absorbed (W) when first stage is at 70°K					
Second stage	35	90	175	225	280
First stage	225	230	275	320	370
Power absorbed (W) at second stage when power absorbed at first stage is zero	50	110	180	220	260

When the power absorbed by the first stage is zero its temperature falls to $42-48^\circ$K, depending on the conditions at the second stage. In the limiting case where no power is absorbed by either stage the temperature of the second stage falls to 12°K.*

Used for the condensation of hydrogen at approximately atmospheric pressure, the Philips PEH-100 Cryogenerator can deliver just under 5 litres-hr^{-1}. Its use as part of the recently introduced Philips helium liquefier is described below. Its requirements are

Electrical power   11 kW
Cooling water   0.75 m^3-hr^{-1}

The Norelco Cryogem Refrigerators†

North American Philips have developed a series of refrigerators working on the same principle but reduced in scale by a factor of about five and adapted to the needs of airborne electronic and

---

Recent development. A two-stage cryogenerator known as the X20 has been developed. It has an integral motor consuming 1.75 kW, and since the mechanical parts are dry-lubricated, the machine can work in any attitude. It will absorb 10 W at 20°K and 30 W at 70°K simultaneously, and has an ultimate second-stage temperature of approximately 12°K.

† See additional footnote on p. 73.

infrared equipment. The range includes both single expansion chamber types, with the cold part extended and insulated so that ultimate temperatures of 35°K can be reached, and also types with two expansion chambers and two regenerators (the regenerators are incorporated in the displacers). The specification of a representative model (the 42160 — see Fig. 3.14) is as follows:

Refrigeration load   2 W at 30°K
Ultimate temperature (zero load)   ~20°K

Fig. 3.14. Model 42160 Norelco *Cryogem* refrigerator. (Photo: Norelco Ltd.)

Power requirement	500 W
Cool-down time	< 12 min
Weight	12 lb including motor
Maintenance	every 250 hr.

The Norelco Micro-Cryogem

The same company have developed still smaller reverse Stirling cycle machines: a single-stage type absorbing 1.5 W at 77°K and a two-stage type absorbing 0.5 W at 22°K. The input power requirement is 35-60 W, the weight including the driver motor is 4-8 lb, depending on the type of motor used, and the overall dimensions are approximately 5½ × 2¾ × 7⅛ in. high.

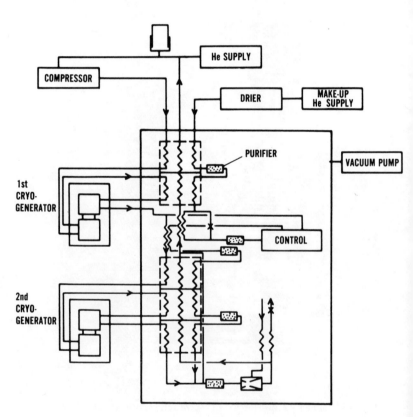

Fig. 3.15. Flow diagram of Philips helium liquefier.

The Philips PL He 209 Helium Liquefier

In 1967 Philips of Eindhoven introduced a helium liquefier which includes several interesting departures from conventionality. Precooling is by means of two type PEH 100 cryogenerators. Helium gas at 20 atm and a flow rate of 60 $m^3 \cdot hr^{-1}$ passes down a chain of heat exchangers (see Fig. 3.15). Between these heat exchangers the helium gas passes successively through both stages of the two cryogenerators, as follows:

	Input temperature (°K)	Output temperature (°K)
Cryogenerator A: first stage	120	100
Cryogenerator B: first stage	75	66
Cryogenerator A: second stage	33	28
Cryogenerator B: second stage	18	16

It will be noticed that there are three paths in the heat exchangers: one is for the stream of helium gas just mentioned, one is a low-pressure path for the returning cold helium, and the third takes a quantity of warm helium gas equivalent to the liquid produced by the liquefier, i.e., the make-up gas. This stream does not pass through the cryogenerators, but joins the stream which does where shown in the diagram. This helium gas comes straight from high-pressure storage cylinders through a pressure-reducing valve and represents about 10% of the gas passing through the liquefier. In a conventional liquefier the high-pressure helium stream would pass through a final heat exchanger and then to an expansion valve. However, in this liquefier, having passed through a fifth heat exchanger of similar pattern to the four already mentioned, the combined high-pressure streams pass at 6.5°K into an *expansion ejector* described by Reitdijk.[11] This takes in the returning unliquefied fraction at 1 atm. The emerging gas stream at 2.5 atm is split into two approximately equal parts: one goes up the low-pressure path of the five heat exchangers and the other goes down a sixth heat exchanger – dissimilar from the others – to the expansion valve. The liquid/gas stream is taken through one of two transfer tubes to an external vessel, and the vapour returned through a parallel tube to the sixth heat exchanger just mentioned. The main purpose of the expansion ejector is to bring the pressure of the returning helium stream up to 2.5 atm since the special oil-free compressor requires this pressure at the inlet to its first stage. The higher pressure of the return gas also eases the heat-exchanger design problems. Further, the pressure over the liquid helium can be close to atmospheric.

The heat exchangers are of a novel design described by Vonk,[12] and Philips will be making them available as individual items. They consist of a stack forming an approximate cube of about 3 in. side made up from sheets of copper gauze interleaved with sheets of a resin-impregnated material in which are punched rectangular holes. The stack is bonded under pressure with an adhesive to form a solid block with channels of varying dimensions across which are stretched at close intervals areas of the copper gauze. Figure 3.16 shows a typical separator sheet in which there are 25 channels. At each end of the stack a header is attached which admits gas to the appropriate channels in the block. Figure 3.17 shows one header and part of a block for a typical two-pass heat exchanger. This is a bold and ingenious design. It is remarkable that such a construction will withstand cooling to low temperatures, quite apart from its being possible to use it where the pressure difference between neighbouring passes is of the order of 20 atm.

The compressor – described in Chapter 4 – is of a novel oil-free design. The liquefier can safely be left running unattended – it shuts itself down in the event of failure of electrical power or cooling water, blockage, etc. A gasholder is not supplied with the liquefier, and at the time of writing no purifying equipment for a recovery system is available.

Specification

Output	8–9 litres-hr^{-1} delivered
Start-up time	2.5–3 hr
Running time between cleaner regenerations	100 hr
Warm-up and regeneration time	4–5 hr
Total weight of all components	3500 kg

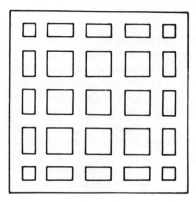

Fig. 3.16. Philips heat exchanger. Typical separator. (Drawing: Philips, Eindhoven.)

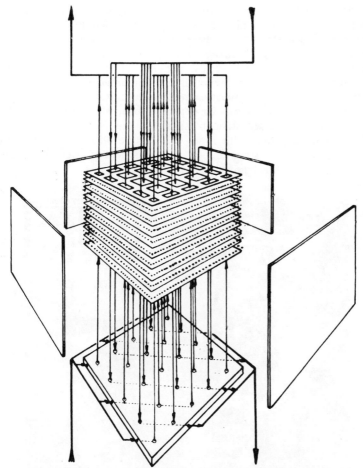

Fig. 3.17. Philips heat exchanger. Header with part of stack. (Drawing: Philips, Eindhoven.)

Requirements

Power	28 kW
Cooling water	2 m^3-hr^{-1} (7.5 gal-min^{-1})
Space	100 ft^2.

The liquefier with the two cryogenerators is shown in Fig. 3.18. At the time of writing five of these liquefiers have been in use for less than a year, and it is therefore too soon to be able to comment on their long-term performance. Among their advantages are that they require no liquid nitrogen and modest electrical power. Their reliability is likely to be determined largely by that of the cryogenerators, which, as mentioned above, perform well if maintained carefully. Where thermal insulation is achieved as it is here by the use of high vacuum there is always the possibility of leakage from one of the many joints, and such leaks can be large enough at low temperature to jeopardize the liquefier's performance, but small enough at room temperature to be difficult to find even with a mass spectrometer. An organization with the reputation enjoyed by Philips can be expected to be able to build a helium liquefier which will not develop leaks, but this is a matter which can only become apparent with time. One feature which might be a disadvantage in some situations is the considerable weight. Another innovation is that this equipment is available on hire.

Fig. 3.18. Model PL He 209 Philips helium liquefier. (Photo: Philips, Eindhoven.)

The Malaker Cryomite Refrigerator

Malaker Laboratories Inc.[Ma] have developed a small-scale reverse Stirling cycle refrigerator called the *Cryomite* which sustains loads of a few watts at temperatures down to $25°K$ using helium gas as the working substance. They are small and light and, being designed for aerospace applications, are normally powered from a 400-cps supply. They are available in the following forms:

Model	Approx. dimensions (in.)	Weight (lb)	Temperature range (°K)	Power absorbed (W)
VII B	13 X 5	13	35–150	1.5 at 40°K
				9.0 at 77°K
VII C	16 X 5	14.5	25–150	5.0 at 30°K
				15 at 77°K

There is also the model VII H, in which the cold head is at the end of a flexible extension, and the model IX, which is adapted for the condensing and recirculating of nitrogen at the rate of 125 g-hr^{-1}.

The ADL Cryodyne Liquefiers and Refrigerators

The Cryodyne cycle which has been developed by Gifford and McMahon[13] of Arthur D. Little, Inc. is somewhat similar in principle to the Kirk (reverse Stirling) refrigeration cycle except that no work is absorbed at low temperature. This led to the use of the misleading description "no-work" cycle. Clearly, if no work is done by the working substance, there will be no cooling: here work is not absorbed *mechanically* at low temperature, but rather at room temperature. A simple example of another refrigerative expansion process in which work is done but not in an engine at low temperature is the Simon helium liquefier (Simon[14]; also see Croft[14]). In this a quantity of helium at, say, 100 atm and 11°K, is expanded to atmospheric pressure. The result of this adiabatic expansion happens – owing to the unusual thermal properties of helium – to result in there being about 60% of the container's volume left filled with liquid helium when the pressure is 1 atm. Work has been done – in this case, against atmospheric pressure. This simple device remains a possible source of liquid helium for small-scale work where a supply of liquid hydrogen is available, and provides a further illustration of the similarity in principle between the use of solid and fluid agents for transmitting the absorption of work from low temperature to room temperature.

The principle of the Gifford–McMahon cycle is as follows (see Fig. 3.19). On the right is a regenerator column capable of working

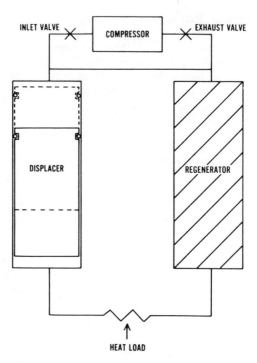

Fig. 3.19. Basic components of Gifford–
McMahon cycle.

between room temperature and 15°K. On the left is a cylinder in which a displacer can be moved. This has a seal at the top and has low axial thermal conductance, so that the bottom can be at low temperature while the top is at room temperature. The tops of the regenerator and the cylinder are joined by a pipe, and so are the bottom ends. There is therefore at no time any net force on the displacer, whose function is merely to move gas from one part of the system to another (cf. the displacer in the Philips machine). With the displacer as shown, at the bottom of its stroke, compressed helium is let in to the system through the external inlet valve. This compresses the helium already in the regenerator and in the space above the displacer, with consequent warming. When the inlet valve has shut the displacer rises and the helium passes through the regenerator into the bottom part of the cylinder, picking up on its way the cold left behind after the previous cycle. The exhaust valve then opens and the helium cools by doing work on the gas on the other side of the exhaust valve. When the exhaust valve has shut, the displacer falls again and the cold gas passes through the regenerator, where its cold is deposited ready for the next cycle. During these last two stages cold gas passes along the pipe joining the cylinder to the regenerator, and this can be used to absorb heat if the cycle is used for refrigeration, or to precool a secondary circuit for a helium liquefier. A typical repetition rate is 72 cycles/min. The thermodynamics of this cycle are complicated because during the inlet and exhaust phases the various batches of gas go through different cycles (see Gifford[15]). However, it is possible to see how a small parcel of the gas can be looked upon as going through an approximation to the reverse Stirling cycle. Let us consider only a parcel of gas above the displacer at the beginning of the cycle described above. This is compressed – approximately isothermally – as in the a to b path in Fig. 3.10a. Still considering only our original parcel of gas, i.e., ignoring the rest of the gas above the displacer and all of that in the regenerator, we transport it intact through the regenerator as the displacer rises, thereby following path b to c. When the exhaust valve opens our parcel of gas

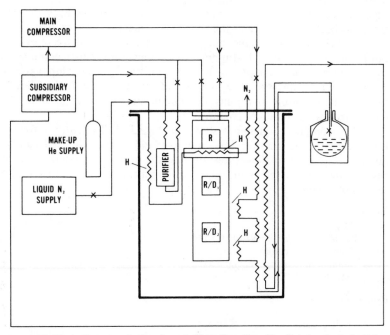

Fig. 3.20. Flow diagram of Cryodyne helium liquefier. H – Points where heat is absorbed.

expands approximately isothermally – the part of the low-temperature piston being played by the gas already in the regenerator. After expansion the parcel is returned to its original state by again being transferred intact through the regenerator. This is a wholly unreal conception: it merely serves to justify the comparison with the reverse Stirling cycle.

This principle has been applied in the very elegant Cryodyne refrigerator which is shown diagrammatically in Fig. 3.20 as applied to a helium liquefier. It has two coaxial displacers, each of which embodies a regenerator (R/D_1 and R/D_2) and also a third regenerator (R). Helium gas at 250 lb-in^{-2} and at a flow rate of 35 ft-^{3}min^{-1} enters the Gifford–McMahon refrigeration circuit and is cooled first in the regenerator R to 80°K and then by liquid nitrogen to 77°K. Then it passes to the two combined regenerator/displacer units R/D_1 and R/D_2, where it is cooled to 30°K and 15°K, returning through the regenerator to the compressor *via* the exhaust valve, on the compressor side of which the pressure is 100 lb-in^{-2}. About 10% of the helium gas leaving the main compressor is tapped off at room temperature. This stream is cooled successively as shown in the diagram at 77°K, 30°K, and at 15°K and then passes into the final Joule–Thomson-stage heat exchanger. From there it goes to an expansion valve which is located in an external liquid-helium vessel. The returning low-pressure gas passes back through the final heat exchanger and then through a chain of heat exchangers in which it helps to cool the incoming secondary stream. This stream of helium gas at approximately 1 atm and room temperature is then compressed by a small subsidiary compressor to 100 lb-in.$^{-2}$ and joins the input stream to the main compressor. The make-up helium gas is purified at liquid-nitrogen temperature and joins the circuit at the same point as the output from the subsidiary compressor.

The complete liquefier is shown in Fig. 3.21. The entire refrigerator unit with the transfer line and expansion valve attached can be raised and lowered to allow insertion into the neck of an external dewar vessel. The whole system including the compressors is mounted into a semi-mobile unit.

Specification

Output	0.5 litres-hr^{-1}
Running time between regeneration	7 days
Time for regeneration	12 hr
Running time before general overhaul	1000 hr
Total weight	1500 lb

Requirements

Power	5 kW
Cooling water	nil
Liquid nitrogen	2.5 litres-hr^{-1}.
Space	50 in. × 30 in. × 72 in. high.

Fig. 3.21. The ADL Cryodyne helium liquefier. (Photo: Arthur D. Little, Inc.)

As a refrigerator the Cryodyne will absorb 3 W if refrigeration is only required at liquid helium temperature. Alternatively, the following loads can be sustained:

Temperature range (°K)	Load (W)
300–77	30
77–30	6.0
30–15	1.7
15–4.3	1.3
4.3	0.5

This equipment has not been in general use for long enough for reports on it to have reached the author. However, the reputation for long-term reliability of the Cryodyne refrigerator, e.g., in the aerospace field, is well established and there seems every reason to suppose from the design of this liquefier that its reliability should be high.

The Unicam Refrigerator

A refrigerator has recently been developed by Pye-Unicam Ltd.[Py] which embodies two refrigeration engines also working on a slightly modified Kirk cycle. As in the Philips devices, but in contrast with the Cryodyne, work is absorbed at low temperature.

Each engine comprises a diaphragm compressor at room temperature driven by oil on the upper side of the diaphragm, a regenerator, and a piston and cylinder operating at low temperature. This piston is sealed by means of a rolling diaphragm at room temperature and is also powered by, and does work on, oil from the same hydraulic system. The working substance is helium gas at a maximum pressure of 400 lb-in.$^{-2}$.

The cycle is as follows:

1. With the piston down and the volume of gas in the low-temperature cylinder zero − although of course there will be cold gas in the bottom part of the regenerator − the diaphragm of the room-temperature compressor is driven down by oil pressure above it. This compresses the working substance approximately isothermally, although in fact much of it is cooled as it fills the regenerator. This compression stage continues until the volume under the diaphragm has decreased by a factor of about three.

2. The piston then rises while the diaphragm continues to fall so as to transfer all of the room-temperature gas into the top of the regenerator and some of the cold gas at the bottom of the regenerator into the cylinder at constant pressure. At the end of this phase the piston has travelled about 22% of its full stroke.

3a. The first part of the expansion phase takes place with the rising of the piston to its fullest extent and the absorption of work by the hydraulic system.

3b. Further expansion then takes place by allowing the diaphragm of the room-temperature compressor to rise fully; as before work is absorbed by the hydraulic system.

4. The piston then falls fully, so that the low-temperature cylinder volume returns to zero, thereby transferring the coldest gas into the bottom of the regenerator and slightly raising the pressure in it and in the space under the compressor diaphragm – in practice, from 100 to 120 lb-in.$^{-2}$.

The two refrigeration engines work at 75°K and 15°K and have cylinder volumes of 70 cm^3 and 13 cm^3, respectively. They precool a separate stream of helium gas at 300 lb-in.^{2} which passes down a chain of heat exchangers to an expansion valve and then back up the heat exchangers. The helium gas is compressed in two diaphragm compressors in three stages, the first two being combined in one unit which raises the pressure to 60 lb-in.$^{-2}$. A separate unit then boosts the pressure to 300 lb-in.$^{-2}$.

The experimental space can be opened up easily. At the time of writing work is in hand on a tailed cryostat for use in magnetic fields and a cryostat with optical windows.

The sequences of operation for starting up and for closing down are controlled by a programmer: pressing a button is all that is required to start up the refrigerator. During operation a preset temperature is maintained automatically.

Specification

Refrigerative capacity	1 W at 4°K
	6 W at 20°K
Start-up time (empty)	4 hr
Dimensions	183 cm $\times$ 70 cm $\times$ 182 cm high
Electrical power required	8 kW three-phase
	and 2 kW single-phase
Cooling water required ($<$ 25°C)	2 gal-min^{-1} peak
	1 gal-min^{-1} mean.

Other External-Work Devices

The following firms have refrigerators or liquefiers in varying stages of development, but insufficient information is available for detailed treatment and the reader is referred to the manufacturers. British Oxygen Cryoproducts[Br1] are working on a liquefier based on an expansion engine in the 5 to 10 litres-hr^{-1} range. The General Electric Company of the USA[Ge] have developed a remarkable refrigerator in which a turbo-compressor, heat exchangers, and turbo-expander are embodied in a cylindrical unit 1 ft diam $\times$ 3 ft long. The Hymatic Engineering Co.[Hy] offer small expansion engines to special order.

Large External-Work Liquefiers

The development of gas-bearing cryogenic turbines – e.g., by Sixsmith and his collaborators[16] – has provided a highly satisfactory means of absorbing work at low temperatures. However, it has so far proved impracticable to apply these to laboratory-scale equipment, i.e., for helium liquefiers producing less than about 50 litres-hr^{-1}. A plant on this scale is beyond the scope of this book, but some of the firms offering equipment embodying turbo-expanders will be found listed in the Appendix. (When further details of the GE refrigerator referred to above become available the statement that small-scale work-absorbing devices of this sort are impracticable may become untrue.)

3.3. COMMERCIAL INTERNAL-WORK LIQUEFIERS AND REFRIGERATORS

Systems in this category are characterized by having no moving parts at low temperature. They depend instead on precooling either by a liquid refrigerant introduced from outside or by one produced by a separate circuit within the system. A refrigeration system working down to 4.2°K could in theory be built using no machinery at all nor any external refrigerants: one could use three high-pressure cylinders filled with nitrogen, hydrogen, and helium. A system like this is extremely inefficient and wasteful, but small-scale refrigerators of this type working down to 4.2°K have been marketed. An internal-work helium liquefier or refrigerator of any size is almost certain to take in liquid nitrogen from outside, and include a liquid-hydrogen precooling circuit involving a compressor, and also, of course, a helium compressor. Usually, it will be the case that vacuum pumps are used to reduce the temperature of the liquid-nitrogen and liquid-hydrogen baths. Comparing this sort of system with one based on external work precooling, we find that we have exchanged the component which absorbs work at low temperature for three pieces of machinery at room temperature, namely, one hydrogen compressor and two vacuum pumps. So far as space goes this is a clear disadvantage, but it is unquestionably the case that room-temperature machinery is more reliable and more easily maintained than highly-specialized low-temperature machinery. In an ideal world the supply of helium gas to a liquefaction system would always be of such high purity that the system would always be capable of removing the residual impurities. However, where recovery systems are in use – and even exceptionally where they are not – the time is certain to come sooner or later when impurity gets into the system. Different types of system vary in their susceptibility to derangement by solid impurities, but one point is certain: an expander of whatever type is

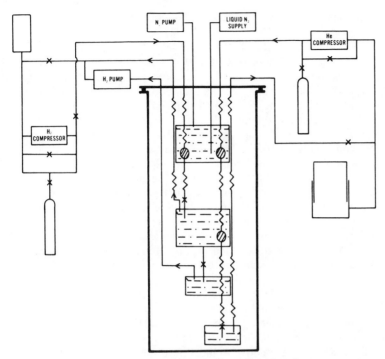

Fig. 3.22. Flow diagram of TBT helium liquefier.

liable to serious damage, whereas an internal-work system, having no moving parts, merely blocks up. Once it has warmed up – whether rapidly by means of special arrangements for introducing warm gas or slowly by normal heat leakage – it can be purged and started up again at once. There is also often an advantage in lower capital cost. An internal-work system may therefore be a competitor with the commoner external-work systems.

The TBT Liquefiers and Refrigerators

The only manufacturer of other than small-scale internal-work liquefiers and refrigerators is the Société d'Études et de Construction d'Appareillages pour les Très Basses Temperatures de Sassenage,* commonly known as TBT[Tr]. There are two helium liquefiers of different capacities from which liquid hydrogen can be withdrawn if

*Now part of the L'Air Liquide[Ai1] organization.

desired, and also a larger hydrogen liquefier. The helium liquefiers are based on the design of Lacaze and Weil.[17]

The circuit of the TBT 7 litres-hr^{-1} (LM7) helium liquefiers is shown in Fig. 3.22. Hydrogen at 120 atm is cooled in a bath of liquid nitrogen maintained at 65°K by a vacuum pump. From here it passes through the final exchanger, and expansion takes place into a liquid reservoir at 22°K. The returning low-pressure hydrogen passes up the final heat exchanger and also through a heat exchanger between room temperature and 65°K. Helium at 25 atm is cooled to 65°K, then to 22°K, and finally to 16°K in a tank which is fed from the tank at 22°K through a valve and maintained at about 120 mm Hg by a second vacuum pump. The helium then passes through its final exchanger to an expansion valve, and the liquid formed is collected in a reservoir. The returning low-pressure helium passes up a chain of four heat exchangers. The hydrogen and helium streams are purified by means of vessels containing charcoal at 65°K and the helium stream by means of a third purifier at 22°K. It will be noted that no attempt is made to salvage the cold from the nitrogen and hydrogen evaporating from the two pumped baths. This leads to a reduction in thermal efficiency, but can be justified by the desirability of avoiding the pressure drop which the use of a heat exchanger would entail.

The heat exchangers are of the classical Linde type: a number of tubes inside a helically-coiled outer tube. This construction suffers from the disadvantage of relatively high pressure drop. For instance, the pressure over the liquid helium is 1.5 atm abs (it is 1.2 atm abs in the ADL–Collins liquefier). This results in a relatively greater loss on transfer to an external vessel. The small volume of the internal liquid-helium reservoir (2.65 litres) used to be a drawback, but this is not so now that the liquefier is available with an automatic draw-off valve. Insulation is by means of a high-vacuum jacket. This makes for compactness, but does multiply the possibilities of trouble from leaks, although reports indicate that these are very rare. Simple automatic control is provided of the level in the liquid-nitrogen and liquid-hydrogen tanks and of the two expansion valves. The electrical equipment is in pressurized enclosures. (See Chapter 9.)

Specifications. Models LM-7A and LM-7 Helium/hydrogen liquefiers

Output of liquid helium	7 litres-hr^{-1}
Output of liquid hydrogen	
(when not liquefying helium)	14 litres-hr^{-1}
Liquid-nitrogen consumption	20 litres-hr^{-1}.

By the close of 1966 TBT had installed 13 hydrogen liquefiers and 35 helium liquefiers, including one and three, respectively, in Great Britain. Their reliability is reported to be good. One advantage of this type of helium liquefier is that liquid hydrogen can be withdrawn and used to precool cryostats — the economic advantages of this have been made clear on p. 8. However, the use of hydrogen in other than a closed circuit calls for still greater precautions (see Chapter 9) and the economics of observing these have to be taken into account.

The Air-Products Cryo-Tip Refrigerator

Air Products and Chemicals, Inc.[Ai2] have produced a refrigerator working in the temperature range 20–200°K which requires only a supply of nitrogen and hydrogen gas from cylinders – there is no provision for conserving these gases, and normally they are exhausted to atmosphere. It will absorb 4 W at 23°K and, if a vacuum pump is used, the temperature can be reduced to 15°K. Temperature-control equipment is fitted, giving a stability of ±0.1°K from 15 to 32°K and ±0.3°K between 32 and 200°K. The refrigerator will run for several hours before the pressure in the cylinder falls to a value at which a full one has to be substituted. Purifying systems are incorporated, but for long runs when commercial grade gas is being used extra cleaners can be added. This refrigerator is available built into a rotatable cryostat dewar fitted with optical windows, or alternatively it can be built into the customer's dewar.

This device is likely to appeal only where liquid nitrogen is not available. Where it is obtainable cheaply a very simple device embodying only one heat exchanger can be made up in the laboratory (see below).

The Hymatic Minicoolers

The Hymatic Engineering Co.[Hy1] have developed a series of miniature internal-work cooling devices. These were originally designed in collaboration with the Royal Radar Establishment, Great Malvern, for military applications.

The simplest Minicoolers are for liquid air, oxygen, nitrogen, or argon and comprise a Collins-type heat exchanger made from cupro-nickel tubing on to which a copper fin has been wound on edge and soldered. Most of them have a fixed orifice instead of an expansion valve, and the three current production types are as follows:

| Model | o.d. (in.) | Overall length (in.) | Max. refrigeration (W) | |
			1500 lb-in.$^{-2}$	4000 lb-in.$^{-2}$
MAC 216	0.187	1.83	4.0 $(0.4\ \text{ft}^3\text{-min}^{-1})*$	10.0 $(0.6\ \text{ft}^3\text{-min}^{-1})*$
MC 8	0.283	1.73		
MAC 217	0.627	2.90	8.0 $(0.6\ \text{ft}^3\text{-min}^{-1})*$	25.0 $(1.6\ \text{ft}^3\text{-min}^{-1})*$

*Flow rate with air

These Minicoolers are supplied with a control unit incorporating a final purifier containing molecular sieve and with a suitable glass dewar vessel. Cool-down time depends on gas pressure, but is of the order of 30 sec.

The model MAC 222 is similar to the MC 8, but is fitted with an automatic expansion valve sensitive to temperature in place of the fixed orifice. This has two advantages: the gas consumption is adjusted to meet variations in load, and the cool-down time is faster because more gas passes through the expansion valve.

Two devices are available for use down to liquid-hydrogen temperatures. The model MAC 215 embodies the Minicooler model MAC 216 to provide liquid nitrogen precooling for a second Minicooler using hydrogen gas. The maximum refrigeration at 21°K is 1.5 W. The model SES 237 uses external liquid nitrogen for precooling and will give up to 4 W of refrigeration at 21°K. Cool-down time is about 15 min.

3.4. LABORATORY-BUILT LIQUEFIERS AND REFRIGERATORS

Now that liquid refrigerants can be purchased in many parts of the world, and with the equipment described above available, it might be thought that it would be as absurd for a laboratory to consider building a liquefier as it would be to make a vacuum pump. However, where there is a shortage of money but no lack of technical competence there are some cases in which it is worth considering laboratory-built equipment. External-work systems can be more or less ruled out, although there have been some exceptional successes – notably, of course, Kapitza's[18] remarkable pioneer liquefier at Cambridge. Refrigeration calling for a few watts at temperatures in the range 20–30°K can be achieved by a very simple system well within the reach of the amateur. If running costs are relatively unimportant, hydrogen can be used straight from a cylinder, as in the case of the Cryotip refrigerator already described, and if cheap liquid nitrogen is available only one heat exchanger is required – for small-scale apparatus this can be of very simple design. A more sophisticated system could include a further heat exchanger to recover the cold of the returning hydrogen. Where there are objections to the use of hydrogen, neon is the alternative working substance, but in this case a closed system including a small compressor (Chapter 4) is necessary. All the information required for the design of simple small-scale heat exchangers will be found in White,[19] Chapter 3.

Refrigerators and liquefiers working down to liquid-helium temperature are naturally more complicated, but nevertheless small-scale apparatus embodying a closed circuit of hydrogen for precooling is not difficult to make. As the scale increases the problems of heat-exchanger design multiply, although the availability of commercial finned tubing – notably that manufactured by Yorkshire Imperial Metals Ltd.[Yo] in the UK and Calumet and Hecla Inc.[Ca] in the USA – has brought the fabrication of larger heat exchangers well within the range of a laboratory workshop (see, e.g., Croft and Tebby[20] and Cosier and Croft[21]). For a bibliography of laboratory-built helium liquefiers see Croft[22] and White,[19] pp. 22-31, and for hydrogen liquefiers Scott et al.[4]

3.5. THE HARWELL DILUTION REFRIGERATOR

Until the light isotope of helium – ^{3}He – became available, the lowest temperature attainable by means of a liquid bath was about 0.9°K. Liquid helium-3 can be used to provide a heat sink down to about 0.3°K, at which its vapour pressure is 1.5 μ Hg. All experimental work at lower temperatures had to make use of the magnetic cooling method. The dilution refrigerator provides a heat sink at temperatures

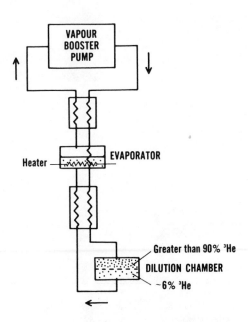

Fig. 3.23. Principle of the ^{3}He dilution refrigerator.

down to about 0.01°K, with the consequences that a new temperature range is available without recourse to magnetic cooling, including nuclear orientation experiments, and also that magnetic cooling experiments can be carried out from a much lower starting temperature — this considerably extends the temperature range accessible in single-stage demagnetization experiments. The commercial availability of the Harwell refrigerator therefore greatly facilitates work at ultra-low temperatures.

In 1951 London[23] suggested that since the entropy of liquid helium-4 is very low at temperatures below 1°K, a cooling effect would result from the dilution of liquid helium-3 by liquid helium-4 analogous to the adiabatic expansion of a gas. This idea remained only a theoretical means of carrying out a discontinuous process until 1956, when Walters and Fairbanks[24] made the unexpected discovery that below 0.87°K mixtures of liquid helium-3 and liquid helium-4 separate into two distinct phases. This property was exploited in 1961 by London *et al.*[25] in the design of a continuous refrigeration system. Refrigerators

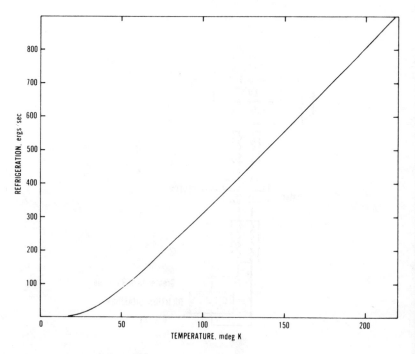

Fig. 3.24. Performance curve of one of the Oxford Instrument Co.'s dilution refrigerators.

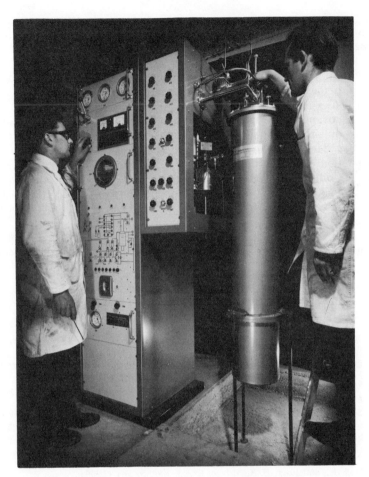

Fig. 3.25. The Oxford Instrument Co., Ltd. dilution refrigerator. (Photo:
Oxford Instrument Co., Ltd.)

embodying this principle have been developed by Das et al.,[26] Hall et
al.,[27] Neganov et al.,[28] and Vilches and Wheatley.[29] The design of Hall
et al.[27] has been developed commercially by the Oxford Instrument
Company Ltd.(Ox1) in cooperation with AERE Harwell. The following
description applies specifically to this variant of the device, although the
principles are the same for the others.

Figure 3.23 shows a closed system which contains a mixture of
^{3}He and ^{4}He. Starting at the exhaust side of the vapour booster pump,

we have ^{3}He gas at a pressure of 30 mm Hg. This passes through a conventional liquid ^{4}He cryostat, where it is cooled to 1.3°K. It then passes through a pressure-reducing capillary, in which it is cooled to 0.6°K at a pressure of less than 1 mm Hg by heat exchange with the evaporator. Condensed liquid ^{3}He then passes through a heat exchanger and into the dilution chamber. This contains the two distinct equilibrium phases referred to above of the ^{3}He–^{4}He solution, the concentrated, almost-pure ^{3}He phase floating on top of the superfluid dilute phase containing 6% ^{3}He. There is a continuous diffusion of ^{3}He across this boundary, with an absorption of heat analogous to latent heat of evaporation. A tube joins the dilution chamber with the evaporator and the ^{3}He diffuses through the more or less static ^{4}He. In the evaporator the ^{3}He is boiled off and passes at a pressure of 0.02 mm Hg to the vapour booster pump. In practice, about 20% of the gas passing through the pump is ^{4}He.

    The curve in Fig. 3.24 shows the performance of a typical refrigerator. Figure 3.25 shows the general appearance of the equipment.

REFERENCES

1. M. Davies, *The Physical Principles of Gas Liquefaction and Low Temperature Rectification,* Longmans Green & Co., London and New York, p. 24.
2. Collins, S. C., and Cannaday, R. L., *Expansion Machines for Low Temperature Processes,* Oxford University Press, London and New York (1958).
3. Mann, D. B., Bjorklund, W. R., Macinko, J., and Hiza, M. J., *Advan. Cryog. Eng.* **5**, 346 (1959).
4. Scott, R. B., Denton, W. H., and Nicholls, C. M., *Technology and Uses of Liquid Hydrogen,* Pergamon, London and New York (1964); see also Singleton, A. H., Lapin, A., and Wenzel, L. A., *Advan. Cryog. Eng.* **13**, 409 (1968).
5. Collins, S. C., *Rev. Sci. Instr.* **18**, 157 (1947).
6. Doll, R., and Eder, F. X., *Advan. Cryog. Eng.* **9**, 561 (1964).
7. Doll, R., and Eder, F. X., *Kältetechnik* **16**, 5 (1964).
8. Köhler, J. W., and Jonkers, C. O., *Philips Tech. Rev.,* **16**, 69, 105 (1954).
9. Van der Ster, J., and Köhler, J. W. L., *Philips Tech. Rev.* **20**, 177 (1958).
10. Prast, G., *Cryogenics* **3**, 156 (1963).
11. Rietdijk, J. A., *Philips Research Papers* (May 1966).
12. Vonk, G., *Advan. Cryog. Eng.* **13**, 582 (1968).
13. Gifford, W. E., and McMahon, H. O., *Advan. Cryog. Eng.* **5**, 354 (1959); see also *Prog. in Cryog.,* **3**, 51 (1961).
14. Simon, F. E., *Z. Phys.* See also Croft, A.J., *J. Sci. Inst.* **29**, 388 (1952).
15. Gifford, W.E., *Advan. Cryog. Eng.* **11**, 152 (1965).
16. Birmingham, B. W., Sixsmith, H., and Wilson, W. A., *Advan. Cryog. Eng.* **7**, 30 (1961).
17. Lacaze, A., and Weil, L., *Proc. IXth Conference of Refrigeration, Paris, Commission I,* (1955), p. 1.

18. Kapitza, P., *Proc. Roy. Soc.* **A147** 189 (1934).
19. White, G. K., *Experimental Techniques in Low-Temperature Physics*, 2nd ed., Oxford University Press (Clarendon Press), (1968).
20. Croft, A. J., and Tebby, P.B., in preparation for submission to *Cryogenics* (1970).
21. Cosier, J., and Croft, A. J., in preparation for submission to *Cryogenics* (1970).
22. Croft, A. J., *Prog. in Cryog.*, 3, 20 (1961).
23. London, H., *Proc. Int. Conf.* Low Temp. Phys., Intern. Union Pure Appl. Phys. *Oxford*, ed. R. Bowers, (1951).
24. Walters, G. K., and Fairbanks, W. M., *Phys. Rev.* **103**, 262 (1956).
25. London, H., Clarke, G. R., and Mendoza, E., *Phys. Rev.* **128**, 1992 (1962).
26. Das, P., De Bruyn Ouboter, R., and Takonis, K. W., *Low Temperature Physics LT9*, Plenum, Press, New York (1965), Part B, p. 1253.
27. Hall, H. E., Ford, P. S., and Thompson, K., *Cryogenics* **6**, 80 (1966).
28. Neganov, B., Berisov, N., and Liburg, M., *Zh. Eksperim. i Teor. Fiz.* **50**, 1445 (1966).
29. Vilches, O. E., and Wheatley, J. C., *Phys. Lett.* **24A**, 440 (1967).

Notes added in proof:

1. The Cryogem refrigerators were developed for military purposes in which lightness and dry lubrication were of greater importance than long-term reliability. They are therefore of limited application in laboratory work.

2. Cryogenic Technology Inc. (formerly Arthur D. Little Inc.) will be delivering early in 1970 the first batch of the new Model 1400 helium liquefier/refrigerator. It has two expansion engines and embodies some of the techniques used in the Cryodyne. Its output is 5-20 litre-h^{-1} of liquid helium or 20-50 W of refrigeration at 4.5°K depending on whether one or two compressor sets is used and whether liquid nitrogen is used. Operation is automatic and the optional built-in purification arrangements allow an impurity of up to 20% of air in the incoming helium gas supply.

3. Philips are introducing a helium liquefier/refrigerator to give 3 litre-h^{-1} of liquid helium or 7 W of refrigeration at 4.5°K. It uses one two-stage cryogenerator and does not require a supply of liquid nitrogen except for the purification system. The compressor used is an adapted Freon refrigerator type having a delivery pressure of 15 atm.

Chapter 4

Room-Temperature Machinery

4.1. COMPRESSORS

Compressors are required in cryogenic refrigeration systems for two distinct purposes: as part of the thermodynamic cycle of a liquefier or refrigerator, and as a means of storing a gas at room temperature; sometimes these functions can be performed by the same machine. The thermodynamic necessity for a compressor has already been made clear by implication in the last chapter. It is not merely there to push the working substance round the system, but in an internal-work system to raise the pressure of the working substance to a value such that at the precooling temperature the enthalpy is smaller than that at atmospheric pressure to the extent required to give the desired performance. In an external-work system the function of the compressor is to reduce the entropy — in this role Simon referred to compressors as "entropy squeezers."

Compressors for cryogenic applications need not differ in principle from those commonly used for other purposes, but certain special considerations may arise, and these are as follows:

1. Conventional compressors are lubricated with oil, and this and any cracking products have to be removed to very low residual levels. This is by no means such a difficult matter as is sometimes supposed, and will be discussed in the next chapter. However, it is clearly desirable that a compressor should not pass an excessive amount of oil into the gas stream and that a long period of use should elapse before wear results in an undue increase in oil transmission. These are both functions of the quality of the design and manufacture. This is one example of a general principle that it pays hands down to buy the best machinery possible. Several types of compressor are available in which the gas is not contaminated with oil, and these will be discussed later. However, it

must be emphasized at this point that such compressors are inevitably more expensive than conventional ones, and may have other disadvantages as well.

2. When drawing up the specification for a compressor one must state the gas with which it is to be used. The temperature on compression varies as $r^{\gamma-1}$, where r is the compression ratio, and is therefore greater for the monatomic gases helium and neon than for polyatomic ones. The designer will take this into account when deciding the number of stages, compression ratio, and shaft speed. In practice compressor manufacturers allow compression ratios of 5–6 for helium (e.g., two stages for compression to about 30 atm) and 7–8 for diatomic gases such as hydrogen (e.g., two stages for compression to about 60 atm).

3. Another reason for specifying the gas is the variation in viscosity between gases. Hydrogen, for example, has approximately half the viscosity of air or helium, and it has been known for a compressor to give a considerably lower output when used with hydrogen than that given when tested with air.

4. With all except air compressors it is important that atmospheric air should not contaminate the gas stream, and there should therefore be no appreciable sources of leakage on the suction side. Ideally, the supply to the compressor should be at a pressure sufficiently above atmospheric and the connecting piping of sufficiently large cross section that the pressure during the suction stroke does not fall below atmospheric anywhere at any time. In this case small leaks will only lead to loss of gas and not to contamination.

5. In most cases it is undesirable that there should be any appreciable leakage of the compressed gas into the atmosphere: in the case of hydrogen more from safety considerations than from any other, and in the case of neon and helium from economic considerations, notably for the former. In all designs of compressor this requirement involves careful attention to all joints and seals and the use of pressure-relief valves from which any blown-off or leaking gas can be collected. The extent of possible sources of loss depends considerably on the design of the compressor.

Conventional compressors are designed according to two basically different concepts. In the so-called *trunk type* the bottoms of the cylinders and of the pistons are open to the crankcase as in the conventional internal combustion engine. Since there is inevitably some leakage past the pistons, the crankcase has to be in communication with the low-pressure inlet to the compressor. Since there is a necessity for the crankcase to "breathe," there should be a pipe of generous cross section between the crankcase and the inlet pipe to the compressor. Certain parts of the crankcase may therefore present possible sources of leak, such as shaft seals, inspection doors, oil filler and dipstick holes, etc., and these may require modification where a compressor has been originally designed for use with air. However, this type of compressor is suitable for hydrogen and helium, although in the case of neon the minimal leakage rate may be found uneconomically high.

The other basic design resembles the now rarely seen steam engine. The crankshaft and crosshead are in a separate enclosure with which the gas to be compressed is not involved. At the point where the piston rod passes through the bottom of the cylinder there is a gland or stuffing box which usually consists of two seals with a pressure feed of oil between them. Thus any leakage will be of oil to the outside or into the gas stream and not of atmospheric air into the gas stream or *vice versa*. This *crosshead* construction is naturally more costly, but is essential for gases as expensive as neon and desirable for other gases where economic considerations are not pressing.

An important property to look for in any type of compressor is the ease with which the inlet and exhaust valves for each stage can be changed. These valves are almost invariably of a type in which a spring-loaded hardened steel washer seats on two annular surfaces across which the compressed gas passes. Wear, distortion, or foreign bodies may prevent the valve from seating. It is therefore essential to keep a set of spare valves at hand, and in a well-designed machine it should be possible to replace one in a matter of minutes. Valves can usually be restored to their original state by lapping.

The heat of compression may be removed either by water-cooling or by air-cooling. The former is usually found in the better-quality machines and is to be preferred, since there is then no difficulty in keeping the machine cool enough nor of disposing of excessive quantities of heat dissipated into the room in which it operates. It will be clear from what has been said above that an air-cooled compressor designed for air is likely to overheat if used for helium or neon; there is usually enough margin in a water-cooled compressor, although manufacturers sometimes recommend running at a lower speed or a lower pressure or both. The amount of cooling water should be adjusted so that the average temperature of the cylinder block is about 60°C. A preset valve in series with an open/shut valve may be used, or for a modest sum a thermostatically-controlled mains water valve may be incorporated. We shall see in the following chapter that it is desirable that the outgoing gas from a compressor should be as cool as possible. Some compressors embody an after-cooler such that the gas leaves the compressor at a temperature close to that of the incoming cooling water, but this is by no means always the case, and quite often no after-cooler is fitted at all. There is therefore likely to be a need for a simple form of water/high-pressure gas heat exchanger which can readily be made by enclosing a helix of copper tubing of appropriate dimensions in an annular cylindrical water jacket.

The choice of a driving motor is not straightforward because compressors present a heavy starting load. If adequately-sized bypass valves are fitted to all stages a high-torque squirrel-cage motor fitted with a star-delta starter is suitable up to 25 h.p. at least provided that a transient current surge of up to three times full-load running current can be tolerated. More satisfactory though more expensive alternatives are the inclusion of an hydraulic coupling, e.g., those manufactured by Crofts[Cr1], or a slip-ring motor with an appropriate starter.

Compressors vary in the extent to which ancillary components are provided with the machine. Pressure-relief valves are invariably fitted, but their outlets may require connecting up to the low-pressure side. In doing this one should incorporate some means of detecting leakage across one of the valves. It should not be necessary to warn readers to check that each stage has the correct relief valve, but having received a compressor from a famous manufacturer on which all four stages had the relief valves appropriate to another, the author feels impelled to do so in view of the dire possible consequences. Gauges reading the pressure between stages should be fitted if they are not already there, since they provide a useful means of diagnosing faults. Abnormal readings while the compressor is running give obvious clues, but still more useful are the readings when the compressor has been stopped: leakages backwards or forwards can be detected at once and faulty valves identified. Apart from the bypass valves which may be required for starting, blow-down valves should be fitted so that all parts of a compressor can be brought down to atmospheric pressure after it has been stopped. These are sometimes included as a means of draining off accumulated oil in interstage coolers.

Most compressors are potent sources of vibration. Where it is necessary to prevent this from being transmitted through the structure of a building, the compressor and motor on a suitable bed can be installed on antivibration mounts, e.g., those supplied by the Dunlop Rubber Co.[Du] In this case considerable movement of the whole unit is inevitable and flexible connections have to be used for the gas, water, and electrical power. In the case of the high-pressure gas line leaving the compressor it is essential to use one of the flexible high-pressure hoses specially made for the purpose, since any arrangement embodying copper tubing is likely to work-harden and subsequently fracture.

Compressors are usually left to run unattended, and the starter control circuit should include switches wired in series which open if the following conditions occur:

1. Fall in gasholder level to near zero.
2. Fall in oil pressure.
3. Fall in cooling water flow or, preferably, rise in compressor temperature.

4. Rise in output gas pressure if this is not already taken care of in some other way. (The pressure relief valves fitted to the last stage of a compressor should be regarded as a last resort only, since these valves tend to wear rapidly.)

There are many makers of pressure-sensitive switches, but a particularly wide range is offered by Black Automatic Controls Ltd.[B1] A useful accessory in the electrical circuit is an hour meter to give some indication of maintenance requirements.

When a compressor is installed the following spare parts should be ordered as a matter of course:

1. A set of driving belts, or, if the machine is direct-coupled, any parts of the coupling susceptible to wear.

2. A full set of inlet and exhaust valves.

3. A full set of piston rings.

4. Shaft seals.

5. A set of main and big-end bearing shells.

6. Blow-off springs.

7. Gudgeon pins, etc., as may be appropriate.

8. In the case of unconventional compressors, any parts which the manufacturer recommends.

The maintenance of conventional compressors can be carried out by anyone who is up to tackling a major overhaul of a motorcar engine. The same diagnostic ear is applicable to sources of impending trouble, except that compressor valves make their own noises which have to be learnt. It is impossible to say how long a compressor can be expected to run without attention, but the following very approximate guide may be useful:

1000 hr – replacement of valves (faults may well have shown up before this time); minor attention to bearings; oil change.

5000 hr – replacement of bearings and piston rings.

10,000 hr – lining and reboring of cylinders and regrinding of crankshaft.

In the Clarendon Laboratory over the last twenty years experience with up to four compressors of various sorts at any one time has shown that appropriate annual maintenance can avoid all breakdowns except those due to faults in valves, which can be put right in a few minutes.

Two Representative Compressors

The names of some compressor manufacturers will be found in the Appendix. Here we describe representatives of the trunk and crosshead types with which the author has had good experience over long periods.

Reavell and Co.(Re) make a wide range of compressors. Of particular interest are their two-stage trunk type compressors which range from the CSA 2 which delivers 9.2 m^3-hr^{-1} at 30 atm, to the HCSA 9, which delivers 110 m^3-hr^{-1} at 65 atm (these figures apply to diatomic gases; for helium the makers recommend a reduction in speed of 30–40% if the output pressure is to remain the same). Figure 4.1

Fig. 4.1. Two-stage HCSA 9 compressor by Reavell & Co. Ltd., Ipswich.
(Photo: C.W. Band, Clarendon Laboratory.)

Fig. 4.2. Four-stage model 150/KK/25 helium compressor by Andreas Hofer of Mülheim. (Photo: R. W. Bowl, Clarendon Laboratory.)

shows the HCSA 9 compressor, which supplies the 30 litres-hr^{-1} hydrogen liquefier in the Clarendon Laboratory (see Croft[1]). This compressor runs at 360 rpm and is driven by a 45 h.p. squirrel-cage motor *via* a Crofts hydraulic coupling (the latter are on the other side of the wall for safety reasons). The first stage is lubricated by oil thrown up from the crankcase, which can be topped up through an arrangement embodying two valves which prevent air from getting in or hydrogen from getting out. Oil is introduced into the second stage by means of a separate oil pump driven from the crankshaft. An interstage cooler and an after-cooler are built into the machine. The four valves are easily accessible — they lie under plates located on the tops of the cylinder heads, and each is secured by nuts. The crankcase breather pipe which communicates with the suction line can be seen in the figure. (Although as we have seen in Chapter 3, p. 30, a pressure of about 120 atm is the optimum for a hydrogen system, the same yield can be obtained if the pressure is 65 atm by an increase in flow of 40%. A compressor delivering at 120 atm is a very much more expensive item than a two-stage compressor of 40% greater output, and this is why the Reavell HCSA 9 compressor was chosen.) The CSA types, which will deliver 30 atm when used with helium, are ideal for all types of helium liquefier, and in particular provide a cheaper and more satisfactory alternative to the compressor sets offered with the ADL–Collins helium liquefier.

Andreas Hofer[An] have been specialists in high-pressure equipment for several decades and offer a range of compressors operating up to much higher pressures than are likely to find application in the field of cryogenics. Their most useful range are four-stage types operating up to 150 atm in capacities ranging from 6.5 to 100 m^3-hr^{-1}. Figure 4.2. shows the 150 atm, 25 m^3-hr^{-1} Hofer compressor used in the Clarendon Laboratory helium-recovery system. The four stages are in line in the order second, first, third, and fourth from the crankcase end – the first stage being double-acting. The cylinders are water-jacketed and the interstage cooling coils lie in a trough of water under the bed. The valves are held in place by the threaded bushes which also form the couplings for the suction and delivery pipes of each stage. The oil pump can be seen at the right of the crankcase. Although these compressors may look old-fashioned, their layout is such that access to all parts of the machine is easy. The quality of design and workmanship and the reliability are all very high, and they have the unusual merits of being very quiet in running and of giving rise to very little vibration.

Oil-Free Compressors

Where there are special reasons for not wishing to use an oil-lubricated compressor followed by the oil-extraction arrangements to be described in the next chapter, certain types of oil-free compressor can be used. Compressors are available – e.g., from Andreas Hofer[An] or Corblin et Cie[Co] – in which a flexible metal diaphragm is pushed up into the dished underside of a circular plate by means of a piston and cylinder working on oil which fills the space below the diaphragm. The gas passes through inlet and exhaust valves located in the top of the dished plate. The fundamental disadvantage of these compressors is that no diaphragm material has been found which does not have a limited life – usually of the order of 1000 hr. Changing the diaphragm is a lengthy and not inexpensive business, and the consequences of premature failure are obvious.

Another approach is to use a piston and cylinder which is conventional except that the piston rings are made of some material such as carbon-loaded polytetrafluorethylene. The disadvantages here are that the piston rings have a limited life and that the dust created by their wearing has to be extracted from the gas stream.

A remarkable new design has been developed by Rietdijk et al.[2] of the Philips Research Laboratories at Eindhoven. The underside of the piston is sealed from the oil-lubricated components below it by means of a rubber sleeve which rolls up and down in the space between the bottom part of the piston and the cylinder wall. By means of steps in the lower part of the piston and of the cylinder it is possible to arrange for the volume of the space which is filled with oil to remain substantially constant for all positions of the piston. Further, by allowing communication between the compression space above the piston and the oil the pressure differential across the rolling rubber diaphragm can be kept low. The sleeve is made of polyurethane rubber and tests have proved that it will withstand more than 10^9 cycles under ideal conditions. In practice, annual renewal is recommended. This description is very much simplified, and the reader is referred to the paper quoted for a full treatment of this principle and also for details of a design in which there is a spring-loaded top cap to the piston. Compressors working up to 150 atm in three stages have been built, although full specifications of production models are not available at the time of writing.

Compressors for Helium Gas Recovery Systems

As will be seen in the next chapter, the recovery of helium gas entails the use of a compressor working up to about 150 atm but of relatively modest capacity. The smaller models in the Hofer range mentioned above are eminently suitable for this purpose. Leybold-Heraeus[Le] offer an interesting direct-coupled, air-cooled, three-stage helium compressor having a capacity of 2.6 m^3-hr^{-1} and working up to 150 atm. A range of air-cooled compressors going down to 1.5 m^3-hr^{-1} is offered by Luchard et Cie[Lu]. Where a compressor working up to about 30 atm forms part of a helium liquefaction or refrigeration system an alternative is to use a booster compressor which has this input pressure and an output pressure of 120–150 atm, thus forming an isolated third stage. (See Appendix for representative makers.)

4.2. VACUUM PUMPS

Although most cryogenic laboratory work involves the use of high-vacuum equipment, this will not be dealt with here because of the wealth of literature on the subject and the many well-known firms offering a wide range of equipment. This section is concerned only with pumps used for reducing the pressure over baths of liquid refrigerants in order to obtain lower temperatures.

A wide variety of pumps is offered by both industrial and scientific engineering firms, and it is possible to save a good deal of money by taking care not to buy a more sophisticated type of pump than the circumstances demand.

Rotary Exhausters

A simple and relatively cheap type of pump is typified by the range of rotary exhausters manufactured by Broom and Wade Ltd.[Br2] The pump is directly coupled to a 1450-rpm motor, and on account of this high speed the dimensions are relatively small compared with the more elaborate pumps described below. An oil pump provides lubrication to the bearings and to the barrel — the oil is not recovered. The blades are of a material similar to that used for brake linings and are caused to follow the contour of the barrel by centrifugal force. The ultimate pressure reached by this type of pump is not lower than 25 mm Hg and the performance of a typical example — the P25 — begins to fall off at about 100 mm Hg. (The manufacturer's value of suction capacity at this pressure is 185 m^3-hr^{-1}, compared with 234 m^3-hr^{-1} at atmospheric pressure. The pump requires a 7 h.p. motor.) This type of pump is therefore suitable, for example, for maintaining a bath of liquid nitrogen at temperatures down to the triple point ($63.2°K$), at which the vapour pressure is 94 mm Hg.

Rotary Vacuum Pumps

There is also a wide range of pumps which are capable of working down to ultimate pressures of the order of 10^{-2} mm Hg. This is achieved by feeding oil to all places where atmospheric air could leak in to the pump from outside or short-circuiting leaks could occur inside. In the small pumps familiar in the laboratory this is commonly achieved by immersing the whole pump in a box filled with oil, but this is naturally impracticable with large pumps. A typical example of this type of pump is shown in Fig. 4.3. This is one of a range of pumps manufactured by Alley Compressors Ltd.[A1] (formerly by the Pulsometer Company of Reading). Two steel blades are pressed against the inside of the barrel by springs and there is a series of discharge flap valves submerged beneath the oil through which the exhaust gas bubbles on its way to the oil separator. To take a specific example, the 12X/63 pump is driven by a 7 h.p. motor at 300 rpm, and at pressures down to below 10^{-1} mm Hg maintains a displacement which is only 10% less than the value at 1 atm, which is 200 m^3-hr^{-1}.

These pumps are useful where pressures lower than 100 mm Hg are required, and an important by-product of the virtually total exclusion of atmospheric air is that they can be used in closed circuits without the introduction of impurities, e.g., for the hydrogen precooling bath in a helium liquefier. Two of these pumps have been in use in the Clarendon Laboratory for 12 and 15 years, respectively, and have given excellent service. After 2000 hr of use minor attention to the blades,

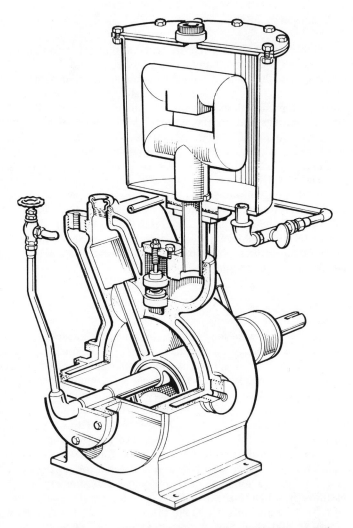

Fig. 4.3. Rotary vacuum pump by Alley Compressors Ltd.
(Drawing: Alley Compressors Ltd.)

springs, and exhaust valve is required, as is renewal of the shaft oil seal and of the oil itself. After very long periods reboring can be carried out, and the construction is such that this can be done several times. These pumps therefore come near to being everlasting.

A pump of similar characteristics but of radically different design is the Kinney[Ki] — this is said to have been an adaptation of a water pump made in the USA when wartime work created a demand for large vacuum pumps. These pumps are manufactured in Great Britain by General Engineering (Radcliffe) Ltd.(Ge2) and in smaller sizes by Edwards High Vacuum Ltd.(Ed) The construction is shown in Fig. 4.4. A cylinder rotates eccentrically inside a barrel and the exhaust valve is a

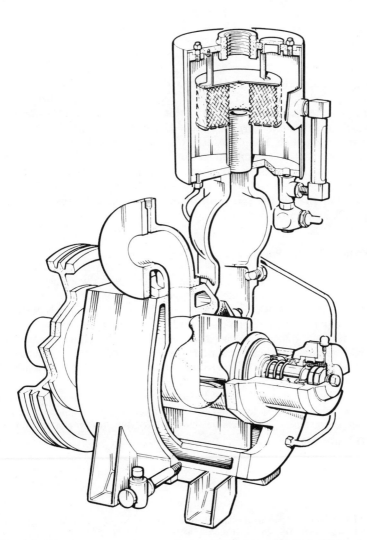

Fig. 4.4. Kinney vacuum pump by General Engineering Co. (Radcliffe) Ltd.

TABLE X

| Type of pump | Pumping speed (litres-μsec^{-1}) | | Displacement of backing pump (litres-min^{-1}) | Power consumption (kW) |
	at 1 mm Hg	at 10^{-2} mm Hg		
9B1	20×10^3	200	450	1.8
9B3	20×10^3	6×10^3	450	2.2
18B3	100×10^3	10×10^3	2,800	6.4

spring-loaded steel strip. However, the cylinder has attached to it a rectangular box such that the cross section of the moving part is like an inverted keyhole. The top part of the box is enclosed in a chamber to which the suction port is attached. The pumped gas passes into the chamber through the open top of the rectangular box and into the barrel through communicating slots in the rotor. This curious design results in a pump of excellent performance, although it must be admitted that its inherent rotational imbalance does lead to more vibration than is the case with other types. These pumps are produced in sizes ranging from 1 to 45 h.p.

Vapour Booster Pumps

There is sometimes a requirement for a pump working to pressures well below 10^{-1} mm Hg, e.g., to maintain a bath of liquid helium at a temperature below 1°K. (The vapour pressure of ^{4}He at 0.9°K is 42μ Hg.) The vapour booster pump – already mentioned as a component in the Harwell dilution refrigerator – is suitable for this pressure range. Many makers of high-vacuum equipment supply them. The examples given in Table X are taken from the range offered by Edwards High Vacuum Ltd.

There is a type of mechanical pump usually known as a Roots pump which is a descendent of the supercharger or "blower" used on high-performance motorcars. This is suitable for pressures down to 10^{-3} mm Hg.

REFERENCES

1. Croft, A. J., *Cryogenics* **4**, 143 (1964).
2. Rietdijk, J. A., van Beukering, H. C. J., van der Aa, H. H. M, and Meijer, R. J., *Philips Tech. Rev.* **26**, 10 (1965).

Chapter 5

Helium Gas Recovery Systems

Only small users of liquid helium – or those blessed with more money than they know what to do with – can afford to let helium out into the atmosphere. Where liquid helium is bought out, it is a matter of returning the helium gas in cylinders to the supplier. There may be an economic advantage in returning the gas at a low level of impurity, and in this case what is said below about purification will be relevant. The owner of a liquefier will want to recover the helium gas evaporated from the cryostats supplied from the liquefier, purify it to an extent acceptable to the liquefier, and store it under pressure at room temperature. In the case of a closed refrigeration system, storage of the charge of gas at room temperature when the plant is not running is likely to be the only problem.

5.1. GAS STORAGE AT ATMOSPHERIC PRESSURE

Where helium is recovered from experimental apparatus a storage volume close to atmospheric pressure is necessary. It may be of importance that the pressure presented should be independent of the contents of the gas container, and it will be seen that this consideration will determine the choice of type, and have a considerable influence on its cost. The unit commonly used to measure pressures close to atmospheric is the inch of water (1 in. w.g. = 1.87 mm Hg = 2.47 $\times$ 10^{-3} atm). Another consideration which has to be borne in mind when choosing a gas container is the permissible loss of helium gas into the atmosphere and conversely of contamination of the helium by atmospheric air. The size will usually be determined by the maximum quantity of gas likely to be returned to the container between times when it is convenient for someone to run a compressor to transfer the gas to storage cylinders. However, economics, shortage of space, or both may be on the side of automating compressor operation (see Chapter 6) and using a smaller gas container.

Fluid-Sealed Gasholders

The simplest and in many ways the best container for low-pressure gas is the old-fashioned gasholder: a cylindrical tank containing water with a counterbalanced and guided bell moving within it. The pressure can be adjusted to some extent and the variation of pressure with contents is very small. This type of gasholder can be calibrated accurately and can then be used in the calibration of flowmeters. The use of water as the sealing fluid is virtually unavoidable for outdoor gasholders, but contamination with water can be avoided by floating a thick layer of oil on top of the water under the bell. Indoor gasholders can be sealed entirely with oil, in which case it is usual to make the bottom container of annular cross section.

There is of course little problem in the case of hydrogen and helium gasholders with corrosion of the internal surfaces. So far as external surfaces are concerned, it is well worth paying extra for a sprayed-zinc coating, which should be further protected by a suitable paint. The protection of the underside requires careful consideration. A slightly-domed concrete plinth should be laid. This should be covered with several layers of thick roofing felt on which a layer of mastic is spread before the bottom part of the gasholder is lowered. With additional sealing round the edge there will then be little tendency for water to creep in nor space for it to lie if it does. A number of manufacturers of traditional gasholders will be found listed in the Appendix, although any competent sheet-metal fabricator should be able to produce one.

The Wiggins Dry-Seal Gasholder

In this type of gasholder, made by the Power-Gas Corporation,(Po) synthetic rubber socks are used to do the sealing. There is thus no contamination at all, and, moreover, the capacity of the gasholder is approximately equal to its maximum external volume rather than to half of this as in the case of the fluid-sealed gasholder. These gasholders are ideal from all technical points of view, but they are relatively expensive.

Gasbags

A much cheaper alternative to traditional gasholders of the types just described is synthetic rubber or plastic gasbags. For example, the

TABLE XI

Permeability of Gasbag Materials*

Material	Temp. (°C)	Permeability		
		N_2	H_2	He
Natural rubber	25	8.2	52	30
Neoprene	20–25	1.2	13.6	537
Polyethylene:				
Low-density	30	–	12	6.5
High-density	30	–	3.0	1.4
Polyvinyl chloride	20–34	0.11	2.4	4.0

* These figures extracted from *Polymer Handbook*.[1]

Dunlop Rubber Co.[Du] offer a neoprene-coated nylon fabric reinforced butyl rubber gasbag 11 ft square which holds 200 ft^3 of gas at a pressure of 4 in. w.g. It may be a disadvantage that these pillow-shaped gasbags do not present a pressure which is independent of content. This is less true of the concertina-like construction, but these are naturally more expensive. The disadvantage of varying pressure can be overcome by means of a low-pressure – in the above case, say, 5 in. w.g. – relief valve such as the bubbler referred to below. The outward diffusion of gas will not be sufficiently great to matter if the best material is used. A value for the permeability to helium gas of a white butyl rubber reinforced with nylon and coated with a weather-resistant neoprene is $3.4 \pm 0.7 \times 10^{-10}$ cm^3(NTP)-sec^{-1} per cm^2 area per cm thickness for 10 mm Hg pressure differential. This corresponds to a loss of about 3 cm^3 of helium gas per day from a gasbag 11 ft square at 4 in. w.g. pressure. This value was obtained in the course of joint experiments between the Dunlop Rubber Co. and the British Oxygen Co. Values for other rubbers in the same units will be found in Table XI. Arthur D. Little, Inc. offer gasbags for helium from 100 to 1000 ft^3 capacity. Other firms manufacturing synthetic rubber gasbags are listed in the Appendix. Although the figure for polythene in Table XI suggests that it ought to be a suitable material for gasbags, the figures for the loss rate quoted by the only manufacturer of these known to the writer are very high. Plastics should therefore be considered as an alternative to synthetic rubbers only if satisfactory data on leakage rate is offered.

There are sometimes applications for portable atmospheric-pressure gas containers holding about 1 m^3, e.g., where small quantities of liquid helium are in use within easy reach of a recovery system but too far away to make a pipeline an economic proposition. The various sizes of meteorological balloons are useful here, although they are very easily torn (see Appendix for suppliers). There is often a need in the laboratory for a gas container holding up to about 5 litres. Football bladders are very useful for this purpose. For example, when a liquid refrigerant is being transferred an inflated football bladder provides not only a suitable pressure, i.e., about 1 lb-in.$^{-2}$, but also provides a useful visual guide to the presence and approximate magnitude of this pressure.

The Non-return Bubbler

A helium recovery system — however small — is bound to include a number of valves to which storage vessels, cryostats, etc., are attached. If one of these is accidentally left open, a considerable quantity of helium may escape. The use of some non-return device is therefore essential, and in a large installation several may be required so as to limit the quantity of helium which can be lost by a single mischance to that which is being returned from a part of the whole system only. Many firms offer suitable valves based on rubber flaps, etc., and the

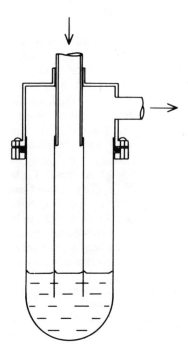

Fig. 5.1. Nonreturn bubbler for low-pressure helium-recovery pipeline.

pressure required to open these and the resistance to flow is adequately low. However, a device which has been found very useful is the oil-filled bubbler illustrated in Fig. 5.1. The tube sizes are chosen to suit the maximum flow. The quantity of oil must then be chosen so that, when the upstream side is open to atmosphere and oil pushed up the central glass tube by the pressure in the gasholder, the oil level in the outer vessel does not fall below the end of the inner tube. By choosing a sufficiently generous diameter for this outer vessel one can keep the additional constant line pressure small. The superiority of this device lies in that an open valve on the upstream side can be detected at once by the rise in oil level up the central tube. Further, the bubbler acts as a crude flowmeter: the evaporation from a storage vessel, for example, can be checked approximately by how often a bubble of helium gas glugs through the oil. The bubbler needs modifying if there is a need to be able to suck gas through it: a volume of adequate size must then be included in the line leading to the central tube to stop the oil from getting into the pipework.

5.2. GAS STORAGE AT HIGH PRESSURE

In laboratories where liquid helium is purchased and the helium gas is recovered and returned to the supplier for credit the problem of storage is merely that of having sufficient standard helium cylinders to contain the gas evaporated between deliveries of the liquid helium. The current standard cylinders which are 9 in. o.d. by 62 in. overall height and which weigh 140 lb hold 260 ft^3 at 165 atm which is the equivalent of 10 litres of liquid helium.

Where a liquefaction plant is in use it is necessary to provide for the storage of at least the quantity of helium gas used in a complete liquefaction run. It is sometimes necessary to have a separate high-pressure store so that contaminated helium can be dumped there pending purification — the necessity for this depends of course on the relationship between the quantity of helium in use and the storage capacity at atmospheric pressure. In the Clarendon Laboratory we have found it essential to duplicate the high-pressure storage for purified helium. One bank of cylinders is filled with purified recovered helium gas, and once this has been tested for purity (see Chapter 6) it can be used for a liquefaction run without fear of trouble from blockages. In the meanwhile the other bank is being filled with purified helium. Should the level of impurity rise owing to some act of carelessness in the laboratory and should this not be spotted in time to prevent, contaminated helium from reaching the high-pressure store, then at least the liquefaction run in progress at the time is not vitiated and all that has to be done is to repurify the contaminated helium gas. High-pressure gas storage cylinders are available in various sizes, e.g., from Chesterfield Tube Co.,[Ch] whose largest standard cylinder holds the equivalent of about 75 litres of liquid helium at 120 atm.

Where a liquefaction system embodies a compressor working at about 30 atm it is possible to save capital expense if use can be made of this in the helium gas recovery system. There are two alternatives: one can either use a booster compressor (see Chapter 4) and use the high-pressure storage cylinders just described, or, alternatively, one can store the gas at pressures up to the maximum of which the liquefier compressor is capable. In this case boiler manufacturers, e.g., John Thompson Ltd.,[Jo2] are likely to offer suitable pressure vessels, or for smaller volumes vessels for liquid carbon dioxide or liquid hydrocarbons may be suitable.

5.3. PURIFICATION

Let us assume that we have a stream of helium gas emerging from a compressor and containing all the impurities which might occur in practice, namely, water, oil, oil cracking products, air, neon, and hydrogen. The following explanation of how these are removed can be applied to other cases.

Liquid water and oil can be removed in a droplet separation column, but it is important first to cool the gas stream to as low a temperature as may be convenient and economic so that as much as possible of the water content is condensed and the vapour pressures of water and oil kept low. As already mentioned in Chapter 4, this may necessitate a subsidiary after-cooler. A typical droplet separator for a compressor of about 50 m^3-hr^{-1} capacity could consist of a vertical pressure vessel – similar to the storage cylinders just described but with openings at both ends – approximately 6 ft tall by 8 in. diameter. The gas stream is introduced at the bottom and let out from the top. The ideal packing for such a column is short pieces of thin-walled tube, say ¼ in. diameter by $^3/_8$ in. long. These are known as Lessing rings and are available as relatively cheap pressings from such chemical engineering suppliers as Visco Ltd.[Vi] A less satisfactory alternative is the use of stone chippings. The principle of such a column is to provide as much surface as possible for droplets to conglomerate upon and to provide a cross section sufficiently great for the upward velocity of the gas not to discourage the downward flow under gravity of the liquid. The accumulated liquid is drawn off from the bottom through a needle valve.

Oil mist and vapour, cracking products, and water vapour can be removed by means of an adsorptive material at room temperature. Of those available the writer prefers Activated Alumina available in Great Britain from Peter Spence Ltd.[Pe] (see Appendix for other suppliers). In purifying columns carrying high-pressure gas there is no pressure-drop problem, and the 16/32 particle size is best. (This grading means that the particles will pass through a gauze filter having 16 strands per in. but not through one having 32). This material will reduce water content to a frost point of about $-60°C$ (water content 0.01 $g\text{-}m^{-3}$). The quantity of water which can be taken up before there is a rise in the water content of the emerging gas stream is about 2.5% w/w of alumina. For still lower frost points molecular-sieve adsorbents made by the Linde Division of the Union Carbide Corp.[Un1] can be used, but for most cryogenic purposes the cheaper alumina materials are good enough. Alumina which has adsorbed oil mist and vapour cannot be regenerated and must be discarded, but adsorbed water can be driven off by passing a stream of heated dry air through the column until its temperature is about 200°C. The direction of flow should be opposite to that when the column is in use so as to prevent any adsorbed oil vapour at the bottom of the column from getting any higher. A drying column reaches saturation without detectable increase in frost point, and it is therefore necessary to keep a spare one ready so that it can be substituted at short notice.

Nitrogen and oxygen are adsorbed at liquid-nitrogen temperature and the best material is probably still activated charcoal. A suitable type is Ultrasorb SC II made by British Carbo-Norit Union Ltd.[Br2] (see Appendix for other suppliers). In the absence of data one cannot give more than the following very approximate working rule: a charcoal purifier of 1 litre volume will extract to sufficiently low levels of impurity about 100 g of nitrogen, oxygen, etc. Experience in the Clarendon Laboratory suggests that this is a somewhat conservative figure. A low-temperature purifier will also remove the remaining trace of water vapour, and reactivation at about 200°C is required to remove this. Adsorbed nitrogen and oxygen are released when the purifier is allowed to rise from liquid-nitrogen temperature to room temperature. A feature of charcoal is that it has a tendency to break up if the gas pressure surrounding it is reduced rapidly. Even with care some fragmentation is unavoidable; fine wire gauze filters should be provided and the charcoal changed at suitable intervals – in the Clarendon Laboratory liquefiers this is done every two years. The partial pressure of nitrogen and oxygen in the gas emerging from such a purifier is greatly dependent on temperature. In a hydrogen liquefier, for example, it is good policy to have a large purifier working at 77°K and a smaller one at 63–65°K if the design of the liquefier allows. A complete cleaning system for helium which can readily be made up in the laboratory is described below.

Neon and hydrogen can be removed by means of activated charcoal at liquid-hydrogen temperature. Hydrogen can also be removed from helium by means of platinum catalyst (e.g., grade D available from Engelhard Industries Ltd.,[En]) if there is either sufficient oxygen there fortuitously or if an appropriate amount is introduced. The resulting water must be removed with an additional drying column. A recording thermometer can be attached to the pressure vessel containing the platinum catalyst to give an indication of how often it has had to perform its duty and thus to give a guide to how much water has been adsorbed by the succeeding drying column.

Components working on these principles are embodied in all liquefaction and refrigeration systems. We now consider specifically equipment for purifying helium gas as part of a recovery system.

The ADL Helium Repurifiers

Arthur D. Little, Inc. offer two units for purifying helium gas embodying charcoal adsorbers working at liquid-nitrogen temperature (Fig. 5.2). The gas streams entering and leaving the charcoal container pass through a countercurrent heat exchanger. The incoming side of this

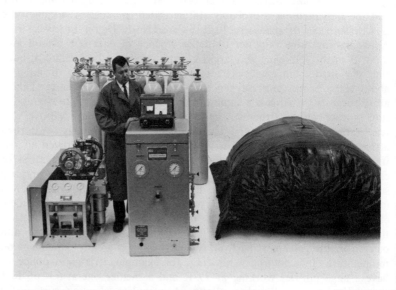

Fig. 5.2. The ADL Model 25 helium purifier. (Photo: Arthur D. Little, Inc.)

is so constructed that oil and water vapour can be condensed without blocking. A katharometric purity monitor is incorporated with a cutout switch to stop the flow when the impurity level rises above the 50 ppm which the makers quote as the performance to be expected, although it is not made clear what level of impurity in the input gas can be tolerated. When the purifier is saturated it can be regenerated in 2 to 3 hr. Specifications are given below.

Specifications	Capacity	Dimensions	Weight
Model 25	3000 ft^3	26 X 22 X 44 in high	250 lb
Model 50	6000 ft^3	26 X 22 X 44 in high	330 lb

Gasbags of 100–1000 ft^3 capacity are available and 6- or 12-cylinder manifolds are supplied. Compressors working up to 1800 lb-in^{-2} are supplied – for the Model 25 of 5 ft^3-min^{-1} capacity and for the Model 50, 12 ft^3-min^{-1} capacity.

Two other recovery units must be distinguished from the one just described in that they embody no means of removing nitrogen, oxygen, etc., from the helium: they consist merely of a low-pressure container, a compressor, a means of extracting contaminants introduced by the compressor, and water vapour if any, together with the necessary valves, gauges, and safety devices. That offered by British Oxygen Cryoproducts Ltd.([Br1]) includes a gasbag of 100 ft^3 capacity and is available with alternative three-stage air-cooled compressors of 0.5 and 2.0 ft^3-min^{-1} capacity. A room-temperature adsorption column filled with charcoal is provided. The unit, which is compact and mobile, does not include any cylinders for high-pressure storage. It is suitable for a laboratory which buys liquid helium from a supplier whose credit terms for returned gas do not place a premium on purity. The gasbag holds the equivalent of just under 4 litres of liquid helium, and this may not be enough for more than the very small user. Leybold Heraeus offer a unit comprising a 450-litre gasbag and the 2.5 m^3-hr^{-1} compressor mentioned in the last chapter. Three high-pressure storage cylinders are provided which hold a total of 18 m^3 at the maximum pressure of 150 atm. The small size of the gasbag is less of a disadvantage because means are provided of starting the compressor automatically when it is full and of stopping it when it is empty. The cylinders provided will hold the equivalent of 25 litres of liquid helium – provision is made for adding further cylinders. The price quoted for this unit seems disproportionately high compared with that of the compressor, which is reasonably priced. For this reason and because there is no purpose in being able to store compressed helium gas if one cannot remove nitrogen and oxygen from it, the appeal of this unit seems somewhat limited. Even if there is no call for purification, one would be better off purchasing the compressor by itself and making up the rest of the system using standard cylinders.

TBT offer as three independent pieces of equipment: a compressor unit similar in principle to those just described, a high-pressure drying unit, and an adsorptive purifier working at liquid-nitrogen temperature. Where no water can get into the gas stream the first and last of these units used together provide similar facilities to those of the ADL purifier, but where a water-sealed gasholder is used the drying unit would be needed as well. The compressor has a capacity of 5 m^3-hr^{-1} and a maximum pressure of 180 atm. An oil separator is included together with a sintered copper filter. Two high-pressure gas-storage cylinders are provided which together hold 7.2 m^3 of gas at 180 atm. There is a spherical gasbag holding 1 m^3 and the compressor is automatically started and stopped according to the pressure in it. The compressor is also stopped and an alarm sounded when the pressure in the storage cylinders reaches 180 atm. The drying unit is available with either a single column, or a pair so that one can be reactivated while the other is in use. Electric heaters are provided for reactivation and cooling water pipes to enable the column to be cooled down more rapidly afterwards. Four types of liquid nitrogen temperature purifier are available: one for pressures up to 40 atm and three for pressures up to 150 atm. The purity of the emerging gas is better than 30 ppm and the capacities range from 240 to 1280 litres of gaseous impurity at NTP.

A Laboratory-Built Purifier

A purifier to extract air and residual water and oil vapour from a high-pressure helium gas stream which has already passed through a column of alumina can be built cheaply in the laboratory. Figure 5.3 shows the principles. The helium gas enters a heat exchanger which consists of two 12-ft lengths of 1/8-in.-bore, annealed, high-pressure copper tubing which have been tinned, twisted together, and then heated up and bonded together with soft solder. The pair of tubes is then wound into a helix of about 6 in. diameter and an insulating annular cylinder of expanded polyurethane foam (see Chapter 7) cast round it. From the bottom of the input pass of this exchanger the helium passes into a few feet of the same tubing immersed in liquid nitrogen. The 4-litre bottle containing the adsorbent charcoal is of EN 58 B stainless steel (AISI 321) obtainable from Chesterfield Tube Co. A virtually empty space is left at the bottom for the condensation of liquid air on occasions when very impure helium has to be cleaned. This space is made by dropping in short lengths of scrap thin-walled tubing — at the

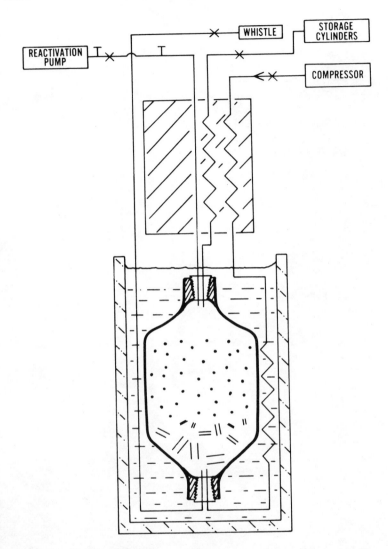

Fig. 5.3. Diagram of typical liquid-nitrogen-temperature adsorption purifier.

bottom the diameter can be as large as will go through the neck, but the top layer must be of sufficiently small diameter to prevent charcoal particles from dropping through. The space can be emptied of condensed air by opening a valve at the room-temperature end of the draining tube leading straight up from the bottom of the bottle. A

Fig. 5.4. Double helium gas purifier in the Clarendon Laboratory. (Photo: R. W. Bowl, Clarendon Laboratory.)

whistle provides a cheap though crude means of distinguishing air from helium. After passing through the bed of charcoal the helium gas enters the other limb of the heat exchanger and emerges close to room temperature. It can be regenerated by means of a sleeve heater and a vacuum pump. A very important safety feature which must on no account be omitted is a relief valve set to a pressure above the normal working pressure but below the safe working pressure of the bottle, so that should the inlet and outlet valves be closed while there is high-pressure gas or adsorbed impurity in the bottle at low temperature or both, a rise in temperature will not lead to disaster. Figure 5.4 shows the helium-purifying equipment in the Clarendon Laboratory; two purifiers based on the design described above will be seen on the left-hand side.

Instruments for measuring purity are described in the next chapter.

REFERENCE

1. *Polymer Handbook*, Interscience, New York (1966).

Chapter 6

Instrumentation

Keeping within the context defined in the preface, we shall be dealing in this chapter only with means of measuring and controlling physical properties concerned with liquefiers, refrigerators, and refrigerant baths in cryostats: the multifarious instruments required for experiments are beyond the scope of this book.

6.1. PRESSURE

The relevant range of pressure is from ~ 10 mm Hg absolute to ~ 150 atm – the measurement of lower pressures belongs to the field of vacuum technology.

Two pieces of usage which may be unfamiliar to physicists must be explained first. The term *gauge* pressure means that the pressure is measured with respect to atmospheric pressure; the term *absolute* means that one is starting from zero pressure and not from atmospheric. It is usual for pressure gauges to read zero at atmospheric pressure and their indication should strictly be written down in the units "atm gauge," or "lb per sq. in. gauge," or the illogical but common "psig." Alternatively, one can add the value of atmospheric pressure *at the time* and record one's readings in terms of "atm absolute," "lb per sq. in. absolute," or "psia." This duality in usage largely arises from and is made more confusing by a fundamental difference between two sorts of dial gauge. The familiar pressure gauge contains a Bourdon tube or capsule stack the inside of which communicates with the pressure to be measured and which is surrounded by atmospheric pressure. This gauge is intrinsically capable only of measuring pressures relative to atmospheric pressure. To take a crude example in engineering terms, consider a simple gauge reading 2.0 lb-in.$^{-2}$ *gauge* pressure. If atmospheric pressure were to change by 1 in. Hg, the reading would change by 0.5 lb-in^{-2}. This is a somewhat extreme example, but illustrates the case that where simple or *noncompensated* gauges are in use the effect of variation in barometric pressure cannot be ignored.*

A more sophisticated type of gauge the readings of which are unaffected by changes of atmospheric pressure is known as a *compensated* gauge. There are two common ways of achieving this. One is to use the differential movement between two pressure-sensitive components, one in communication with the pressure to be measured and the other sealed off. The other is to enclose a single sealed-off pressure-sensitive component in a pressure-tight case to which the pressure to be measured is admitted — this construction is naturally only suitable for relatively low pressures. Such gauges are always calibrated in terms of absolute pressure. Dial vacuum gauges, if they are of the noncompensated sort, read zero at atmospheric pressure and 76 cm Hg or 30 in. Hg at the other end of the scale. *Compound* gauges are those with zero in the centre of the scale, at which the pressure is atmospheric, and scales reading vacuum and pressure *gauge* pressures on either side.

Failure of Bourdon tubes in pressure gauges is not common but does occur, and can have serious consequences if high-pressure gauges are wrongly chosen or used. For pressures above about 50 atm special gauges are available, e.g., the Fig. 13 G.P. range manufactured by the Budenberg Gauge Co.[Bu2] These have a thin copper or plastic sheet at the back, a substantial casting between the casing containing the Bourdon tube and the dial, and laminated glass on the front. This construction affords a high degree of protection for the user — provided that he is on the right side of the gauge when it blows up and that he has not committed the folly of attaching the back of the gauge firmly to a solid panel or wall. Where a pressure gauge is in a part of a circuit close to a compressor there are likely to be rapid fluctuations in pressure which may cause premature fatigue failure of the gauge tube. The remedy is to fit a small needle valve which can be adjusted so that the pointer does not vibrate but still indicates long-term variations in pressure. High-pressure gauges are usually calibrated hydraulically with oil, and there are often generous traces of this remaining when gauges are received from the manufacturer. It should be removed with a suitable solvent unless the gauge is to be used in a part of a system where the gas is contaminated with oil in any case. Manufacturers rarely supply connecting fittings automatically, and time and trouble will be saved if the user remembers to order them with the gauge. Nuts securing the connecting tailpiece to high-pressure gauges have a curious way of working loose, and routine tightening is recommended.

* Indeed, some 15 years ago the operator of an earlier Clarendon Laboratory helium liquefier used to make reliable weather forecasts from the readings of one of the gauges.

Differential gauges have two connections and indicate the difference between the pressure applied to each – the limitation on the so-called static pressure is always clearly indicated on the dial. The Microvar differential gauges manufactured by the Budenberg Gauge Co. will operate at static pressures up to 2000 lb-in.$^{-2}$ and will indicate differential pressures as small as 3 lb-in.$^{-2}$; they are therefore useful for flowmeters for gas at high pressures. Differential gauges for lower pressures can also be used for flow measurement, although it will be seen below that there are alternatives. Differential gauges will also be mentioned in connection with liquid level measurement. As an indication of the sensitivity available in the best type of purely mechanical instrument, Negretti and Zambra Ltd.(Ne) offer a gauge with a 10-in.-diam scale having a full-scale reading of 2 in. water gauge (usually abbreviated to "2 in. w.g.")* at a maximum static pressure of 40 in. w.g.

     A recent development is the pressure-sensitive transistor. These are not competitive in price with mechanical gauges at present, but are likely in the course of time to make a considerable impact in the field of pressure measurement. An example in current production provides an output of 1 V for a pressure differential of 3.5 in w.g. with a 1% linearity and an overload capacity of 500%. This is the Stolab Pitran, manufactured by Stow Laboratories, Inc.(St)

     Gauges of high quality are individually calibrated against a dead-weight tester or a mercury column according to the pressure range, and the makers will give figures for accuracy and will check the calibration after long periods of use. There is the difference between instruments of the highest quality and those next in line that the latter should be tapped before a reading is taken whereas no pointer movement results from tapping the former. There are many makers of gauges which offer advantage in price and delivery, but the writer's small experience of these suggests that they should not be regarded as instruments suitable for scientific use.

     Table XII summarizes some available pressure gauges.

6.2. TEMPERATURE – ELECTRICAL METHODS

The methods of temperature measurement to be described below have been selected for convenience and availability combined with moderate but adequate accuracy (see White,[1] Chapter 4, and Hoare *et al.*[2] for a fuller treatment).

---

* See p. 87 for pressure conversion factors.

TABLE XII

Availability of Some British Pressure Gauges*

Supplier	Special high-pressure gas gauges	Medium–high-, low-pressure and vacuum gauges (uncompensated)	Compensated vacuum gauges	High-pressure differential gauges	Low-pressure differential gauges	High-sensitivity, low-pressure differential gauges
Negretti and Zambra(Ne)	Up to 10,000 lb in.$^{-2}$	Yes	Yes	No	Yes	Yes
Budenberg Gauge(Bu$_2$)	5000 lb in.$^{-2}$ standard, higher to order	Yes down to 0.5 in. w.g.	No	Yes	Yes	No
Cambridge Industrial Instruments Ltd. Ca$_2$	Up to 10,000 lb in.$^{-2}$	Yes down to 1.0 in. w.g.	No	No	Yes	Yes
Edwards High Vacuum(Ed)	No	—	20, 40, 100 760 mm Hg	No	No	No

*Notes. 1. All makers supply pressure gauges calibrated in atmospheres. 2. Higher sensitivities for low gauge and differential pressures are available in servo-operated types (e.g., from Ne and Ca). 3. Some makers offer recording instruments (e.g. Ne, Bu$_2$, Ca). 4. Other manufacturers will be found listed in the Appendix.

Electrical Resistance Thermometers

Thermometers depending upon the resistivity of pure metals are limited in their usefulness by the diminishing resistivity/temperature gradient at low temperatures. Platinum is still the standard material for measurement of temperatures down to about 20°K. A particularly useful form is that manufactured by Degussa[De] in which a fine platinum wire is fused on to a glass rod approximately 20 mm by 3 mm. The room-temperature resistance is 100 ohms and a calibration down to 23°K is supplied. This construction is not ideal because the wire is unlikely to be in the state of freedom from mechanical strain essential for accurate work. White[1] gives makers of less convenient but more accurate forms. The usual way of using resistance thermometers is to attach two wires to each end and use one pair to provide the measuring current and the other to measure the voltage across the thermometer. A variety of alloys can be used for resistance thermometers at temperatures below 20°K, but most of these suffer from a limited useful temperature range, relative insensitivity, and difficult availability. They have been superseded by semiconductor devices and special thermocouples.

A favourite form of resistance thermometer for many years has been the carbon resistor commonly used in electronics. Much work has been done on their reproducibility and calibration (see Bibliography in White[1]). Their use remains something of a black art in that certain makes almost invariably work well, whereas others do not. The Allen Bradley and Speer types have been the most used. Values in the region 50–100 ohms are commonly used. The relationship of resistance with temperature approximates to $\ln R \propto 1/T$. They are commonly used in the range 1–100°K.

The carbon resistance thermometer is a crude form of the semiconductor thermometer, which in recent years has been developed into an instrument of high accuracy and reproducibility but also of relatively high cost. Typical thermometers consist of slices of single-crystal doped

TABLE XIII

| | Extremes of Useful Range | | | | |
| | Min T | | Max T | | Sensitivity |
Series	T (°K)	R (kilohms)	T (°K)	R (ohms)	(ohms-deg K^{-1})
I	1.5	20	40	10	500 at 4.2°K
II	1.5	10	100	5	400 at 4.2°K
III	10	2	100	6	50 at 20°K
IV	1.5	4	10	100	100 at 4.2°K

germanium or silicon usually sealed into a capsule with helium gas to provide thermal exchange. Calibrated thermometers are available at extra cost. A typical range is that offered by Honeywell, Inc.(Ho2) shown in Table XIII. The maximum recommended power dissipation is 2.5 μW, and measuring currents have to be limited accordingly. Other makers will be found listed in the Appendix. These devices are intended for accurate experimental work and will not often be found appropriate for measurements on the equipment dealt with in this book.

Thermocouples

Thermocouples have two advantages for low-temperature measurement: they are small, and since the heat introduced by their presence can be made negligible, there is little difficulty in making sure that the temperature of the thermocouple is accurately that of the object to which it is attached. The traditional copper/constantan thermocouple is useful between room temperature and temperatures down to about $50°K$, over which range its sensitivity falls from ~40 to ~13 μV-deg K^{-1}. However, its sensitivity becomes awkwardly small at temperatures much lower than this, being ~6 μV-deg K^{-1} at $20°K$. Care needs to be taken over selection of the constantan wire because inhomogeneity may give rise to spurious voltages (see White,[1] p. 149). There is a wealth of information in the literature on the calibration of copper/constantan thermocouples (e.g., Powell *et al.*[3] and the National Bureau of Standards[4]). A thermocouple which can be used for accurate measurements between 1° and $300°K$ has been described by Berman *et al.*,[6] who give a full discussion of it and also of papers relating to its genesis and to earlier low-temperature thermocouples. The materials used in this thermocouple are spectroscopically pure gold with 0.03 at. % of added iron (obtainable from Johnson, Matthey & Co. Ltd.(Jo4)) and the common thermocouple alloy *Chromel*, which has the composition 90 Ni 10 Cr (obtainable from Amalgams Co. Ltd.(Am)). The sensitivity between 1° and $300°K$ lies within the limits 10 and 20 μV-deg K^{-1} with two points of inflexion. The reproducibility over two years has been better than 0.5%. This combination therefore provides an ideal thermocouple for the whole of the temperature region with which we are concerned.

6.3. TEMPERATURE – PRESSURE-DEPENDENT METHODS

The electrical methods described above are appropriate where high accuracy, wide range, and small mass or heat influx – or some of these properties – are called for. However, unless sophisticated electronics is used, indication of temperature on a panel-mounted instrument is out of the question, and these methods are generally to be avoided for routine use in liquefiers and refrigerators.

There are two alternatives which indicate temperature on a dial pressure gauge: the vapour-pressure thermometer and the helium gas thermometer. The range of a vapour-pressure thermometer is of course limited to that between the triple point and critical point of the gas with which it is filled. The volume of the thermometer bulb has to be chosen so that over the temperature range it is always partly filled with liquid. Naturally, the dead volume of the gauge and the maximum pressure it will stand have to be taken into account. From what has been said above it will be clear that for temperature readings of any accuracy a compensated gauge is essential. If the gauge manufacturer is given the vapour pressure/temperature curve the gauge can be calibrated directly in terms of temperature and an accuracy greater than $0.1°K$ is easily achieved over a short range. Vapour-pressure thermometers are especially useful to check the temperature of the gas leaving a pre-cooling bath before it enters the final heat-exchanger of a Linde liquefier, since, as we have seen in Chapter 3, the performance depends steeply upon this temperature. Filled and calibrated vapour-pressure thermometers are available commercially, but at such a price that most laboratory workers will prefer to make up their own.

In the traditional gas thermometer the dead volume at room temperature is kept as small as possible. There is a useful form of gas thermometer in which the abandonment of this principle is turned to advantage. Suppose we have a dial vacuum gauge at room temperature

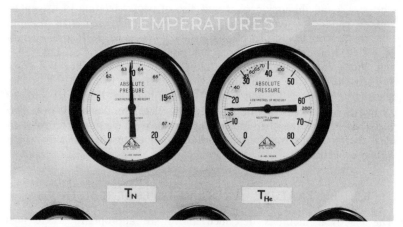

Fig. 6.1. Nitrogen vapour-pressure and helium gas thermometers based on compensated absolute gauges by Negretti & Zambra Ltd. (Photo: R. W. Bowl, Clarendon Laboratory.)

whose dead volume is greater than that of a thermometer bulb at low temperature by a factor r. The two are joined by a capillary tube the volume of which can be neglected. Application of the perfect gas laws shows that if the system is filled with helium gas to atmospheric pressure at room temperature T_0 the gauge will indicate 38 cm Hg, i.e., half scale, when the temperature is $T_0/(r + 2)$. Below this temperature the temperature scale becomes progressively more open; above, it is somewhat cramped. An appropriate choice of r enables one to select a useful range for any specific application. For example, if one is interested in temperatures below $10°K$, one will choose a value for r of about 30; for temperatures below $100°K$ the ratio would be about three – necessitating a size of bulb which might in some circumstances be an embarrassment. Once again, a compensated vacuum gauge is virtually obligatory. Figure 6.1 shows a nitrogen vapour-pressure thermometer reading the precooling temperature before the final heat exchanger and a helium gas thermometer in thermal contact with the expansion valve embodied in the Clarendon Laboratory hydrogen liquefier (see Croft[5]). These are both Negretti and Zambra compensated vacuum gauges. Temperature values have been painted on the glass.

6.4. GAS FLOW RATE

A continuous flow-rate instrument for gas at atmospheric pressure is an essential part of a liquefaction or refrigeration system. It can be used to check the performance of a compressor and, placed in the return stream, to measure the liquefaction rate if the flow rate entering the liquefier is known. A cheap instrument which is sufficiently accurate for most purposes of this sort is a slightly tapered vertical glass tube containing a float which takes up a position corresponding to the flow rate engraved on the glass tube or on a scale attached to it. Oblique flutes on the float cause it to rotate, and this usually stabilizes it. These instruments are available for the largest flow rates likely to be encountered in laboratory systems down to very small ones suitable for checking evaporation rates from cryostats, etc. Naturally, the calibration depends upon the gas and upon its temperature. (See the Appendix for manufacturers.) Figure 6.2 shows a Rotameter calibrated from 5 to 50 m^3-hr^{-1} for helium gas at atmospheric pressure. Alternatively, a Venturi tube or orifice plate can be inserted in the pipeline and a differential pressure gauge used to indicate the flow.

More sophisticated gas flow meters are available which measure mass flow independently of temperature and pressure over a specified range and which indicate on a digital read-out. Meterflow Ltd.(Me) offer an instrument in which the rate of revolution of a turbine and measure-

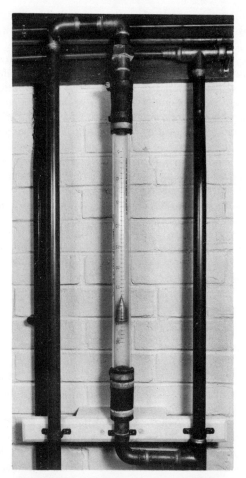

Fig. 6.2. Flowmeter by Rotameter Manufacturing Co. Ltd. (Photo: R. W. Bowl, Clarendon Laboratory.)

ment of the density of a gas are combined electronically to give an indication of mass flow rate. These instruments are available for flows ranging from 0.06 to 10^4 ft^3-min^{-1}. Fischer and Porter Ltd.[Fi1] have recently introduced an instrument called a *Swirlmeter* which has no moving parts and is available for a wide range of flow rates and for pressures up to 2000 lb-in.$^{-2}$. The gas stream, which must already be in the state of turbulent flow, is passed through a fixed set of blades which impart a rotary motion to it. This results in pulsating temperature changes, which are detected by a thermistor and converted electronically into an indication of flow rate. The gas stream is subsequently restored to its original state of flow by a second set of blades. At high and low flow rates the instrument is not linear and the calibration depends upon viscosity and compressibility, but over the linear range the calibration is independent of density, pressure, temperature, and viscosity.

It is often useful to be able to measure gas flows at high pressures, and for this the high-pressure differential gauge made by the Budenberg Gauge Co., which has been mentioned above, can be used across a length of tubing to provide the necessary pressure drop. The length and diameter can be calculated from the expression for pressure drop under turbulent flow conditions given in White[1] on p. 75. Such an instrument can be calibrated by expanding the gas to atmospheric pressure and passing it through one of the instruments described above or through one of the integrating meters described below.

6.5. INTEGRATED GAS FLOW RATE

The domestic gas meter is an example of an instrument which records the volume of gas which has passed through it, i.e., it registers the integral of the flow rate with respect to time. This type of dry gas meter is useful in the laboratory in checking the efficiency of helium conservation systems and for measuring very small flow rates. They do not maintain their calibration indefinitely and require checking against the wet type described below.

For accurate measurement of volumes of gas the wet type of gas meter is to be preferred. A cylindrical drum rotates about a horizontal axis inside a casing approximately half filled with water to a definite level. As gas passes through the instrument the drum is caused to rotate by the successive filling of chambers in it with gas and the displacement of the gas from them by water. Provided that the maximum stated flow rate is not exceeded, the accuracy of this type of instrument can be relied upon. (See the Appendix for manufacturers.)

6.6. LIQUID/VAPOUR INTERFACE POINT

There are two distinct types of level measurement device: those which indicate whether a particular point is above or below the surface of a liquid, and those which provide a continuous indication of depth. The former are the subject of this section.

An ingenious gadget which is applicable to all liquids except liquid helium is the optical dip-stick. A Perspex rod is used as a light guide with the end cut at such an angle that up to 800 times more light is reflected back up the light guide when the end is submerged than when it is not. Where hydrogen is present modification may be necessary to make the device electrically safe (see Chapter 9). (See Appendix for manufacturers.)

A device of even greater simplicity is one based on acoustic principles. It works with – to the novice – startling efficiency with liquid helium, less well with liquid hydrogen and liquid neon, but not at all with higher-boiling refrigerants. A thin-walled, poorly conducting tube, typically about 3 mm o.d., has a funnel attached to its upper end across which is stretched a rubber diaphragm some 3 cm in diameter. With the open end of the tube below the liquid-helium surface oscillations of the rubber diaphragm of 10–20 cps can be felt by the finger. The tube is then raised slowly. As the open end clears the surface the frequency and amplitude of vibration rise sharply. With skill a liquid-helium surface can be located to within a few millimetres. This device boils off liquid helium at the rate of about 2 litres-hr^{-1}, but this is of little consequence since it is quick to use. It was first described by Gaffney and Clement[7]; the physical rationale is given in Hoare *et al.*[2] p. 155.

Most other liquid-surface detectors depend on the combination of a thermometer with a means of introducing heat. The interface can then be detected by the change in temperature resulting from the difference between the thermal exchange between the thermometer and the liquid, and between it and the vapour. A cheap and simple variety embodies a vapour-pressure thermometer in which the capillary tube is made of copper of sufficient cross section to conduct enough heat. The system may be filled with oxygen for detecting the level of liquid nitrogen, or neon for liquid hydrogen; this method is unsuitable for liquid helium. Alternatively, use can be made of the various forms of resistance thermometer already described, although for level-measurement applications one has to find out by experiment what the minimum power is at which the thermometer will work as a liquid-surface detector – this will be very much greater than the power necessary for its use as a straight thermometer.

A doped silicon diode has been produced by the French Firm CSF(CS) under the brand name Silec, which has excellent characteristics for use in liquid-surface detection devices. Current/voltage curves published by the makers for immersion below and just above liquid surfaces of helium, hydrogen, and nitrogen show a pronounced difference in voltage across the diode for an optimum applied constant current. The characteristics vary slightly from one diode to another and the optimum operating point has to be determined experimentally. Binks and Hudson[8] have described a neat instrument using these diodes which detects the level of liquid helium to within 1 mm by the lighting and extinguishing of a lamp. The current through the diode is ~5 mA and the difference in the voltage across it varies from 1.2 to 0.9 V when the probe head is in and out of the liquid, respectively. The probe head embodies the lamp, a switch, a simple transistor circuit, and a small battery from which the current drain is 45 or 15 mA according to whether the lamp is alight or not. These instruments are available from Tilbury Cryo-Equipment.(Ti)

6.7. LIQUID DEPTH

The indication of depth of liquids with densities not far removed from that of water, e.g., liquid nitrogen, is most easily measured hydro-statically. The classical way of doing this is to take capillary tubes from the top and bottom of a tank, to attach them to the two sides of a differential pressure gauge, and – except in the unusual case where the bottom tube goes down rather than up – to include a means of displacing the level of liquid in the capillary tube coming from the bottom of the vessel, usually by passing a very small stream of gas down it. This principle can be applied to cryogenics without the necessity of the latter complication by making the tube leading from the bottom of the vessel of sufficient thermal conductance to ensure that there can be no liquid in it. For example, in the case of a metal dewar vessel a copper capillary can be passed through the outer wall, into and down the insulating space to a point below the bottom of the inner vessel. A piece of tube of low thermal conductance and of, say, 10 cm length can then be used to connect it to the interior of the inner vessel. One can then be sure that the liquid/vapour interface will be in the horizontal, poorly-conducting tube and not in the vertical copper tube. An alternative where it is not possible to connect a tube to the bottom of a vessel is to introduce from the top a copper capillary with a suitable insulating jacket of poor thermal conductance. (Account should be taken in the design of differential thermal contraction.) The application of this principle to liquid helium and liquid hydrogen is less easy because simple differential pressure gauges are not obtainable for ranges of less than 0–2 in. w.g.

Another simple device which depends upon pressure measurement has been described by Biermans and Nihoul.[9] A piece of stainless steel tubing of about 1-mm bore and closed at the bottom occupies the range of possible height of the liquid the level of which is to be measured, and is joined to a pressure gauge with added dead volume at room tempera-ture. The system is filled with the gas the liquid phase of which is the subject of the depth measurement, and the pressure and dead volume so related that the former is always above the pressure of the liquid bath, i.e., the levels inside and out can keep in step. The reading of the pressure gauge is linear with level to a sufficient accuracy for most purposes. The authors quote a range of 850–1050 mm Hg absolute with a dead volume of 350 cm^3 and a room-temperature pressure of 1100 mm Hg absolute. In the case of liquid helium the ratio of the densities of the liquid and the vapour at the boiling point is as low as 7.5, and care needs to be taken to see that enough heat leakage is arranged for the part of the tube containing vapour to be well above the boiling point. The authors suggest a heater, but a more attractive alternative is the use of a copper capillary tube between the top of the stainless steel measuring tube and room temperature. The wall thickness of the measuring tube is also relevant; in the device described it was 0.25 mm. This device is especially suited to level-control applications either by using the varying pressure directly in a flexible metal bellows (see p. 116 below) or as part of a servosystem.

A crude method of continuous level measurement is to make up a chain of carbon resistance thermometers. The sensitivity is clearly limited by the length of the resistors, and the amount of power introduced may be an added disadvantage.

For depth measurement of liquid helium the use of supercon- ducting devices is obviously attractive. Although several alternatives have been put forward, no single technique has as yet become universally adopted. The principle is to choose a material for a wire which becomes superconducting as close as possible to 4.2°K and to pass a large enough current through it to make sure that the wire becomes normally conducting not far above the surface of the liquid. Provided that the resistance of the wire above the liquid can be taken as constant, i.e., that it has a low temperature coefficient of resistivity or that the temperature is constant, then the resistance of the wire is linearly related to the contents of the vessel. An ideal material might be thought to be tantalum, the transition temperature of which is 4.38°K, and this was first used for this purpose by Feldmeier and Serin,[10] but it has not gained popularity on account of the difficulty of obtaining tantalum wire of sufficient purity, and also because in its simple form variations in the temperature of the wire above the liquid surface limit the accuracy.

This latter disadvantage was overcome by Ries and Satterthwaite,[11] who used a manganin wire on which had been electrodeposited a layer about 10 μ thick of a Pb/Sn alloy of 40/60 w/w composition. The transition temperature of the alloy is 4.8°K. Depth-measuring devices of this sort all suffer from the disadvantage that in order to keep the part of the wire above the liquid in the non-superconducting state, sufficient heat has to be dissipated in this part of the wire to give rise to an evaporation rate that cannot always be tolerated if the instrument is in continuous operation.

The methods of surface detection and depth measurement which have been described are those which are at the time of writing most commonly used. The simplest and cheapest of all is the float with an indicator extending upwards. These work satisfactorily for liquid nitrogen and other refrigerants with densities in the neighbourhood of unity, but are unattractive for liquid hydrogen and liquid helium. Capacitance gauges have been used and also sonic methods for detecting liquid surfaces. A full bibliography is given in White.[1]

6.8. GAS PURITY

Methods of removing impurities from gas streams have been dealt with in the last chapter. It is important to be able to monitor the performance of these systems and also to be able to tell in advance from the impurity of the incoming gas how long one can run the purifier before it is likely to become saturated. It is also useful to be able to check the purity of commercially-obtained make-up gas so that one knows whether it can be fed straight into a liquefier or refrigerator without initial purification.

Mass-spectrometer gas analysis equipment is clearly by far the most versatile and sensitive for this purpose, but it is also expensive and may not always be justified by the particular needs concerned. A form of mass spectrometer especially useful for cryogenic applications is one in which continuous indication is provided of the partial pressure of those gases likely to be present – preferably on a logarithmic scale. For example, the helium gas recovery system in the Clarendon Laboratory would find of great value an instrument giving continuous indication of the content of hydrogen, neon, nitrogen, and oxygen in the helium gas passing to the purification system. Manufacturers will be found listed in the Appendix.

Katharometric instruments depend on differences of thermal conductivity of gases. For example, hydrogen and helium have comparable thermal conductivities which are six to seven times those of oxygen and nitrogen. Katharometers are therefore suitable for detecting the presence of either of the latter in either of the former, but not of either of the former in each other. A typical katharometer offered by Electronic Instruments Ltd.[El1] is calibrated from 100 to 96% hydrogen – or helium – on a galvanometer scale 17 cm long on the assumption that the impurity is oxygen or nitrogen or a mixture of the two. Higher sensitivities are available with additional equipment to keep the katharometer block at constant temperature and to amplify the signal. For most systems in which helium or hydrogen is the working substance and air the likely impurity such instruments provide a reliable and relatively cheap means of measuring impurity down to 0.1%.

Where it is required to measure the concentration of hydrogen as an impurity in helium a catalyst instrument can be made up cheaply – no commercial instrument of this sort seems to be available. A thermally-insulated tube about 1 cm i.d. by 10 cm long has a quantity of platinum catalyst occupying a length of a few centimetres at its centre. Semiconducting thermometers or thermistors are embedded in loosely-packed glass wool plugs on either side of the catalyst. The thermistors are included in a bridge circuit such that any difference in temperature is indicated on a meter. This instrument can be used to detect the presence of oxygen in hydrogen or *vice versa* and can be calibrated for a specific flow rate – e.g., indicated by a small Rotameter – by means of gas mixtures of known composition. Application to the measurement of hydrogen in helium necessitates the introduction of a small quantity of oxygen – or air – unless the sample is already known to contain sufficient to react with the whole of the likely hydrogen impurity.

The water content of gases can be determined by means of a number of commercially-available instruments (see Appendix) many of which are based on measuring the change of capacitance between two electrodes enclosing a hygroscopic layer, the dielectric constant of which is greatly altered by very small traces of water. Other types are based on the change of resistance across a hygroscopic solid electrolyte. Hygrometers may be calibrated in terms of either frost point or water content. The values in Table XIV are given to help in choosing an appropriate instrument.

TABLE XIV

Frost point ($^\circ$C)	Water content (parts per million v/v)
−100	0.0130
−80	0.526
−60	10.6
−40	127.0

6.9. AUTOMATIC CONTROL

Temperature stabilization in experimental apparatus is in general beyond the scope of this book, but it seems appropriate here to review the various gadgets which are available to control the pressure over a pumped bath of a liquid refrigerant and thus to keep it at constant temperature. The expensive method of doing this is by means of a servosystem, and the best are probably the air-powered types offered by many manufacturers (see Appendix) and used widely in industry. They incorporate a pressure gauge movement equivalent to more or less any of those described above, i.e., compensated or otherwise, differential or otherwise, etc. The controller is fed with clean, dry compressed air, usually at 20 lb-in.$^{-2}$, and an output signal pressure is delivered which varies, usually between 3 and 15 lb-in.$^{-2}$, according to the difference between the actual dial reading and that of an adjustable pointer. The simpler type of so-called single-term controller has an adjustment which varies the range of difference between the actual reading and the adjustable pointer reading over which the signal pressure varies: this is called the proportional band and is expressed in terms of the percentage of the full-scale reading. With the proportional band set low, control will be more accurate, but the tendency to hunt will be greater. However, for most cryogenic applications a compromise can be arrived at which gives adequate accuracy combined with stability. Where this is not possible more sophisticated controllers are called for, details of which will be found in textbooks on control engineering. The output signal can be fed to a pneumatically-powered valve, of which there are two distinct types. In the simpler the signal pressure is applied to a diaphragm or a piston working in a cylinder and the force balanced against a spring – any force on the valve spindle arising from within the valve is compensated for only as a result of the action of the controller. A more sophisticated type of control is provided by a valve positioner which is an air-powered servosystem resulting in a definite valve opening for any given value of applied signal pressure irrespective of any variation in the force on the valve spindle. (Makers of valves of both types will be found listed in the Appendix.) Such systems can be applied to the automatic control of expansion valves, etc., in liquefiers and refrigerators – see, e.g., Croft and Robertson.[12]

A number of cheaper and simpler but less accurate devices are available commercially or have been devised by experimentalists and can easily be made up in the laboratory. These devices do not in general operate with complete stability under all conditions: a small degree of hunting will occur and the potential user has to decide whether this is acceptable. One such device is the Cartesian manostat, in which the valve is operated by a component similar to the diver in the well-known scientific toy. The controlled pressure can be varied by altering the pressure inside the diver, which moves in a well filled with mercury. The Model 6 Cartesian manostat marketed by Edwards High Vacuum Ltd. has a differential between opening and closing of $\sim 0.5\%$ over the range 100–400 mm Hg; the maximum flow rate at the latter pressure is 4 litres-min^{-1}. The range of controlled pressure over which this manostat will operate is 5–2740 mm Hg absolute. (See Appendix for other manufacturers.) White[1] describes in detail (his p. 245) how a glass Cartesian manostat can be made up in the laboratory, and also a number of other devices based on metal bellows, rubber diaphragms, etc.

Liquid refrigerant level control devices vary from those used in liquefiers and refrigerators to keep a refrigerant bath at a more or less constant level to those used to keep liquid-nitrogen traps topped up. It is usually not a disadvantage if such devices operate discontinuously, and in some cases it is a positive advantage that they should. For example, where a precooling refrigerant bath in a liquefier or a refrigerator operates at reduced pressure the average performance over long periods will be better if the precooling refrigerant is replenished discontinuously – with consequent large rise in pressure for a short period – than if it is replenished at a constant rate. For the control of liquid nitrogen a vapour-pressure system filled with oxygen is simple and reliable. The principle is similar to that mentioned above in connection with the detection of liquid-refrigerant surfaces. A bulb is provided with a heat leak, e.g., by means of a copper tube to connect it to the room-temperature part of the apparatus, and positioned where it is required that the maximum height of the liquid should be. The bulb communicates with a flexible metal bellows which may operate a liquid-refrigerant valve or may cause a storage vessel to be pressurized and so cause the transfer of, say, liquid nitrogen to a condensation trap. The liquid-refrigerant valve of simple construction illustrated in Fig. 6.3 is suitable for most applications, but it should be noted that the supply pressure must not be greater than the pressure in the bellows when the bulb has warmed up divided by the ratio of the effective area of the bellows to the area of the valve seat. The valve spindle is inserted from the bottom and the slot in the plug – in which the tongue on the needle is a very loose fit – is to stop it from becoming unscrewed from the threaded boss on the underside of the bellows. The two connections to the bellows are for the capillary tube leading to the bulb and for the tube through which the working substance is introduced, and which is subsequently sealed off.

Where continuous stable control is required the depth gauge described by Biermans and Nihoul and referred to in Section 6.7 above can be coupled to a pneumatic servosystem and the output signal used to control a liquid-refrigerant valve *via* an actuator or positioner.

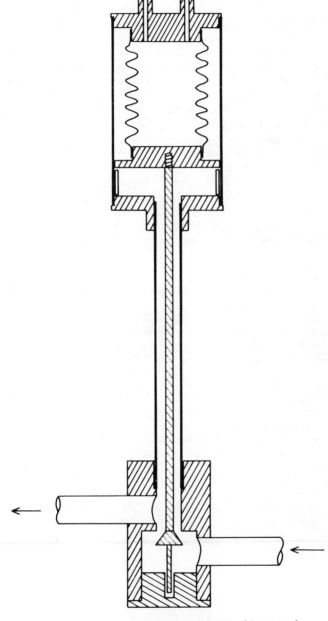

Fig. 6.3. Vapour-pressure-operated liquid-refrigerant valve.

Commercial devices for topping up liquid-nitrogen trap dewars include the following: Spembley Technical Products Ltd.[Sp] offer a system in which a solenoid-operated liquid-refrigerant valve is controlled by a thermistor probe; Harvey Control Co.[Ha] offer a similar device; Oxford Instrument Co.[Ox] offer a device in which a liquid-nitrogen pump is controlled by a temperature-sensing unit embodying a bellows; KDG Instruments Ltd.[KD] offer alternative systems embodying a liquid-refrigerant pump or a valve – either is controlled by a gas-filled tube the pressure in which operates a microswitch. Similar equipment is offered by A. D. Little, Inc.,[Ar] and Gardner Cryogenics Corp.[Ga₁]

The automatic control of gas-recovery compressors can be achieved by means of switches sensitive to gas-container content and solenoid-operated high-pressure valves to let the gas into the high-pressure storage containers and to blow down the compressor when it has stopped. Protective circuitry to guard against overheating, excessive pressures, oil-pressure failure, etc., should of course be fitted in any case, as mentioned in Chapter 4. Components are available from many sources, and a few representative ones will be found listed in the Appendix.

REFERENCES

1. White, G. K., *Experimental Techniques in Low-Temperature Physics*, 2nd ed., Oxford University Press (Clarendon Press) (1968).
2. Hoare, F. E., Jackson, L. C., and Kurti, N., (eds.) *Experimental Cryophysics,* Butterworths, London (1961), Chapter 9.
3. Powell, R. L., Caywood, L. P., Bunch, M. D., *Temperature* 3 (Part 2), 65 (1962).
4. National Bureau of Standards Circular 561 (1955).
5. Croft, A. J., *Cryogenics* 4, 143 (1964).
6. Berman, R., Brock, J. C. F., and Huntley, D. J., *Cryogenics* 4, 233 (1964); Berman, R., and Huntley, D. J., *Cryogenics* 3, 70 (1963).
7. Gaffney, J., and Clement, J. R., *Rev. Sci. Instr.* 26, 620 (1955).
8. Binks, R. A., and Hudson, P. A., in preparation for submission to *Cryogenics* (1969/70).
9. Biermans, F., and Nihoul, J., *Cryogenics* 2, 243 (1962).
10. Feldmeier, J. R., and Serin, B., *Rev. Sci. Instr.* 19, 916 (1948).
11. Ries, R., and Satterthwaite, C. B., *Rev. Sci. Instr.* 35, 762 (1964).
12. Croft, A. J., and Robertson, C. W., *Cryogenics* 9, 365 (1969).

Chapter 7

Materials and Jointing Methods

7.1. DESIRABLE FEATURES OF STRUCTURAL MATERIALS

The construction of low-temperature equipment calls for materials with special physical properties, and the use of many materials at low temperatures brings special problems of its own. The relevance of various physical properties will be discussed first, and then we shall review the extent to which actual materials have them.

Thermal conductivity and its variation with temperature is of importance both when it is a matter of keeping an unwanted heat influx to a minimum and when the maximum heat flow is required. For example, in a cryostat one is concerned to minimize heat flow down neck tubes, etc., and also to maximize the conduction of heat down a radiation shield so that the temperature gradient is minimal.

Tensile strength and other mechanical properties are of great importance in that some metals suffer a change in structure causing them to become brittle at low temperatures. With this exception, it is a general rule that mechanical properties improve with decrease in temperature.

Thermal expansion — generally thought of in terms of contraction in this context — has already been mentioned in Chapter 2 in connection with transfer lines, but has often to be taken into account elsewhere, especially where materials of dissimilar thermal contraction are joined.

Emissivity has also been mentioned in Chapter 2 and will be considered in detail in relation to thermal insulation below. In general, the emissivity of a surface is low for good electrical conductors provided that the thickness is adequate and the surface chemically clean.

A material is said to be of high *integrity* if one can rely on its being what it is supposed to be. For example, in the case of a tube any

119

Representative Values* of Some Properties of Selected Constructional Materials

Material	Heat flow ($J\text{-}sec^{-1}\text{-}cm^{-1}$)		Tensile strength ($kg\text{-}cm^{-2}$)			Contraction $(\Delta l/l)(\times 10^{-3})$	Emissivity (from 300°K)	Enthalpy change ($J\text{-}g^{-1}$)		
	300–76°K	76–4°K	300°K	77°K	20°K	(293–20°K)	(minimum values)	300–70°K	70–20°K	20–4°K
Copper (high purity, annealed)	930	3070	2200	3400	4400	3.4	0.013	75	4.1	0.033
Copper ("high conductivity")†	934	586	—	—	—	—	—	—	—	—
Aluminum	570	1820	4400	5500	7000	4.15	0.011	163	6.1	0.0086
Aluminum alloy (typical)	265	35	4800	5700	6700	—	~0.02	—	—	—
Brass (typical)	156	16	4000	5700	—	3.5	0.018	~80	~3	~0.04
Cupronickel (typical)	88	19	5600	8000	8800	~2	~0.1	—	—	—
Stainless steel (300 series)	27	3.2	6200	19,000	20,000	~2.7	0.05	~80	~2	~0.03
Glass (Borosilicate)	1.8	0.17	—	—	—	0.6	—	—	—	—
Polytetrafluoroethylene (Teflon, Fluon, etc.)	0.57	0.13	140	900	—	21	—	160	9.1	0.52
Methyl methacrylate (Perspex, Lucite, etc.)	0.53	0.020	—	—	—	9.9	—	—	—	—
Nylon	0.76	0.13	630	1900	—	14	—	—	—	—

*These values are intended for guidance only: for reliable data on specific materials see publications listed in the Bibliography.
†Included only for comparison of heat flow.

porosities or inclusions must be of dimensions so much smaller than the wall thickness of the tube that there are no leakage paths. This is a matter of everyday concern at room temperatures, but is aggravated in low-temperature work for several reasons: (1) low thermal conductance calls for very thin-walled tubes, typically down to 0.1 mm, made from alloys such as stainless steel; (2) the physical properties of fluids at low temperatures are such that they can get through small channels more easily than at room temperature, notably the extreme case of superfluid helium; (3) experience shows that leaks which are absent or small at room temperature show up or become larger ar low temperatures — sometimes only after many temperature cycles (see Section 7.7). Care in selection and use of materials is all-important, but no one should expect to escape altogether from the consequences of imperfect integrity.

Magnetic properties can be of importance in the design of cryostats for certain types of experiment. Many metals which are not ferromagnetic at room temperature become so at low temperatures, notably, many nickel alloys. The behaviour of those which do not and of the commonly-used stainless steels is often capricious, being dependent upon slight variations of composition within the specification, and upon the physical state for a given composition. It is therefore advisable when in doubt to do a crude experiment. Difficulties from magnetic fields caused by induced currents in superconductors must also be considered, notably in the case of solders. (See Section 7.5.)

Specific heat is not of importance in the choice of structural materials, since none of those commonly used have any special advantage over others. (The case is of course different in the choice of packing material for regenerators — see p. 46).

Table XV gives representative data for a variety of structural materials, which are intended only to illustrate the variation in the properties just reviewed and for use in approximate calculations. For sources of accurate data on a wide variety of materials see the Bibliography.

7.2. METALS

Copper is an obvious material where high thermal conductivity is required, but as reference to Table XV will show, there is a considerable spread of thermal conductivities over the range of materials commonly so described, and this spread widens at 4°K to a factor of 40. Where the highest thermal conductivity at the lowest temperatures is required, copper of very high purity is therefore necessary, and this may be difficult to get in the form required. One is therefore usually obliged to make do with what is available and either to measure the thermal conductivity of a specimen or to estimate it from the specification. Despite its increased price relative to other metals, and its low mechanical strength, copper remains a favourite material for pipework on account of the ease with which it can be manipulated and jointed and of its high integrity. Further information on copper tubing will be found below. As Table XV shows, the emissivity of high-conductivity copper is adequately low, but this applies only when it is chemically clean. (A polished surface is of no advantage, and may even give a higher value because of the reduction of electrical conductivity by work-hardening.) When a copper radiation shield has to be exposed to the atmosphere frequently its performance can be maintained by gold plating.

Brasses are still useful where there are reasons for wishing to use tin/lead solders, e.g., in low-temperature parts of cryostats which have to be demountable. The choice of brass is important especially where hard soldering is involved, since some brasses contain lead to facilitate machining and this causes the alloy to become porous if subjected to temperatures above about $500°C$. BSS 378, 885, 886, 1402, and 1403 (tubing) and BSS 265 and 267 (sheet and plate) are substantially lead free, but BSS 249 (extruded sections), a "free-cutting" brass, contains up to 4% lead and should be avoided. Blanks should therefore be cut from sheet rather than from bar. Brasses are usually reliable as concerns integrity, the rolled sheets being more reliable* than the extruded bars and sections. (This is a further reason for preferring sheet to bar as the source of material for blanks.)

Aluminium and aluminium alloys are being increasingly used on account of their cheapness, lightness, and mechanical strength, following the advent of reliable inert-gas welding methods. Design must take into account the requirements of welding techniques. The difficulties of joining aluminium and its alloys to stainless steels are discussed in Section 7.5 below.

Copper/nickel alloys were extensively used where low thermal conductivity was required, but they have been largely superseded by the stainless steels mentioned below. These alloys have the advantage that they can be more readily soft-soldered. Alloys which have been popular are Inconel (80 Ni, 14 Cr, 6 Fe), cupronickel (30 Ni, 0.7 Fe, 0.8 Mn, remainder Cu), and german silver (60–65 Cu, 20 Ni, 15–20 Zn).

Low-expansion alloys have an obvious application where thermal contraction is an embarrassment in long transfer lines. An example is the alloy 36 Ni 64 Fe — similar to the long-established Invar — which is available commercially as Nilo-36 from Henry Wiggin and Co. Ltd.[He] and from other suppliers (see Appendix). The contraction ratio between

*See Note 1 on p. 146.

$293°$ and $20°K$ is 0.5×10^{-3}; the values shown in Table XV for other commonly used metals are four to eight times higher. The usefulness of this material is vitiated by the difficulty of obtaining it in tubing of suitable dimensions.

Stainless steels are now so relatively cheap and widely obtainable that their use for the room-temperature parts of cryogenic apparatus is becoming common. The choice of alloy for this purpose is not relevant here; we shall be concerned only with stainless steels used at low temperatures. Their low thermal conductivities coupled with high mechanical strengths make them ideal materials from which to draw thin-walled tubes of minimal thermal conductance. The austenitic stainless steels of the 18/8 type are in general highly suitable for use at low temperatures, since their tensile strength increases, as does their ductility, down to the lowest temperatures. The three alloys most used are as follows: EN58E (A1S1 304) has the composition 0.08 C, 2.00 Mn, 1.00 Sc, 0.045 P, 0.030 S, 18–20 Cr, 8–12 Ni, remainder Fe. In the fully annealed state it is non-magnetic and corrosion-resistant. When cold-worked, i.e., in the case of tubes in the "as-drawn" condition, it is slightly magnetic. Its one disadvantage is that if annealing is not carried out after welding carbide precipitation may occur which vitiates the resistance to corrosion. This is overcome in the following two alloys, which are "stabilized." EN58B (A1S1 321) and EN58F (A1S1 347) have very nearly the same composition as EN58E, but include titanium to the extent of five times the carbon content and niobium to the extent of ten times the carbon content, respectively. When these alloys are subjected to welding temperatures the carbides formed stay in solution and the chromium remains unaffected. Consequently, the corrosion resistance is unaltered even without subsequent annealing. Niobium is slightly to be preferred to titanium as a stabilizer because it does not have the latter's tendency to come out of solution in certain circumstances and so lead to porosity.

Carbon steels must on no account be used at low temperatures, since in some circumstances there may be dangerous consequences from brittle fracture which arises from a transition to martensitic structure. There is no need to treat this matter in detail because there are no components likely to be required for use at low temperatures which are not available in 18/8 stainless steels.

7.3. NON-METALS

Plastics have still lower thermal conductivities than metals (see Table XV) and quite apart from their use in heat insulation (see below), they have obvious attractions as constructional materials, e.g., as the neck tube in a liquid-helium storage vessel. The snag is that they have high coefficients of thermal expansion, and this leads to difficulties with jointing.

   Glass — the original and for many years the almost universal material for dewar vessels — has now virtually disappeared from the cryogenic scene except in certain specialized experimental applications. This being so, attention must be drawn to the once all too familiar porosity of borosilicate glasses to helium. Between a glass such as *Pyrex* and soda-glass there is a factor of 100 in the rate of diffusion, but since the latter is unsuitable for use where wide temperature variations occur, it has been usual to use a compromise between the two such as *Monax* — a glass produced by John Moncrieff and Co.[Jo3] The rate of diffusion D is strongly temperature-dependent: $\ln D \propto 1/T$. If contact with helium can be avoided when glass is warm, troubles arising from diffusion will be minimized.

7.4. TUBING

The choice of dimensions of tubing of any specific material to carry gases at room temperature depends upon the pressure, which determines the relationship between wall thickness and diameter, and upon the tolerable pressure drop, which depends upon the properties of the gas concerned and its flow rate.

TABLE XVI

Ratio of External Diameter/Wall Thickness for Specific Bursting Pressures of Tubes of Selected Materials

Material	External diameter/wall thickness			
	150 atm	300 atm	500 atm	750 atm
Copper (half-hard)	27	23	17	11
Copper (annealed)	18	16	12	7
Cupronickel (annealed)	28	24	21	15
German silver (as-drawn)	65	42	10	—
Stainless steel (annealed)	68	51	*	*

* No tubing available which bursts at these pressures.

Table XVI shows the relationship of *bursting* pressure to outside diameter/wall thickness for many specimens of tubes of various materials. (The author is indebted to Mr. E. R. Tilbury of the Clarendon Laboratory for access to data recorded over many years.) The linear relationship shown by materials in the hard state is absent in those which have been annealed. This is because bursting is preceded by bulging, and the state of work-hardening which is reached before bursting occurs depends on the original thickness of the wall. (In general, bulging occurs at pressures between 30 and 50% of the bursting pressure.) The recommended rule of thumb is to take the maximum working pressure as 0.2 of the bursting pressure. In the case of metals such as copper which anneal at the temperatures required for hard soldering the choice should be based on data for the annealed state when hard-drawn tubing is thus jointed.

Collapsing pressures are mainly of interest in the case of thin-walled stainless steel tubes where the pressure is atmospheric outside the tube and close to zero inside. Use of the nomogram given by Hoare *et al.*,[1] confirmed by experimental determinations, suggests that for annealed stainless steel the ratio of diameter to wall thickness will be just satisfactory at 150, but in practice a value below 75 is recommended.

In Great Britain there are two differing principles on which the sizes of copper tubing are based. (The present situation is confusing enough, and its rationale will become still more obscure as the inch becomes less fashionable.) One system is to settle on a series of standard outside diameters and to offer a variety of wall thicknesses to suit the pressure. These are known as "O. D. sizes" and run from $^1/_{16}$ to $^3/_4$ in. by $^1/_{16}$ in. and up to 4 in. by $^1/_8$ in. The other system originally started with an internal diameter which was a fractional inch size, and an arbitrary wall thickness (BSS 659 introduced in 1936). This resulted in an outside diameter which could only be expressed in terms of decimal inches. In due course the wall thicknesses have been reduced, and the sizes commonly used now have bores which are larger than the original fractional inch sizes; this has led to the use of the term "nominal bore" for the system. The outer and inner diameters of selected sizes conforming to BSS 659/1955 and BSS 659/1967 are given in Table XVII below. The original fractional internal diameters are preserved in BSS 1386 for the range $^1/_8 - ^1/_2$ in., i.e., for these the nominal bore is identical with the actual bore.

This tubing is widely used for domestic plumbing and is therefore readily obtainable. It is often referred to as "Yorkshire," being manufactured by Yorkshire Imperial Metals Ltd., Leeds,[Yo] a subsidiary of Imperial Chemical Industries Ltd. This firm will draw copper and copper alloy tubing to any specification to special order. They also manufacture

TABLE XVII

Dimensions of BSS 659 tubes (in.)

Nominal bore	BSS 659/1955 o.d.	BSS 659/1955 i.d.	BSS 659/1967 i.d.
$^1/_8$	0.205	0.149	0.161
$^1/_4$	0.346	0.274	0.298
$^3/_8$	0.471	0.399	0.419
$^1/_2$	0.596	0.516	0.542
$^3/_4$	0.846	0.766	0.778
1	1.112	1.016	1.040
1½	1.612	1.516	1.524
2	2.128	2.016	2.032

a wide range of fittings for hard- and soft-soldering. In the "capillary" type for soft solder a ring of solder is pressed into a groove in an untinned fitting. The procedure recommended for using these fittings is satisfactory for plumbing work, but cannot be relied upon for the high performance often required in scientific work (see Section 7.5 below).

An alternative to soldered fittings for room-temperature pipework is the compression type such as the widely-used Ermeto[Br4] range. An advantage of this means of jointing is that the cheaper aluminium and aluminium alloy tubing can be used. For low-pressure applications, e.g., helium gas-recovery systems, the use of PVC tubing and fittings has obvious advantages.

Brass tubing is still much used in the construction of low-temperature equipment, since it can readily be hard- or soft-soldered and is available in a wide variety of sizes (see Appendix for suppliers). The range of 41 sizes of Rollet telescopic brass tubing is especially useful to the experimentalist: it runs from letter C, having an o.d. of 0.125 in. and a wall thickness of 0.018 in., to number 38, having an o.d. of 2.116 in. and a wall thickness of 0.039 in. It was introduced by, and is still available from, H. Rollet and Co. Ltd.[Ro3]

Seamless stainless steel tubing drawn to the minimum possible wall thickness is, as has already been made clear, the obvious material for the necks of storage vessels and cryostat dewars, etc. Seamless tubing is available in such alloys as EN58B (AISI 321) and EN58E (AISI 304), in sizes from $^1/_{32}$ in. o.d. by 0.002 in. wall to 1¼ in. o.d. by 0.036 in. wall. (These are the minimum wall thicknesses available.) Larger-diameter tubes are made by rolling sheet and making a seam by fusion butt welding without the use of any filler material. The resulting tube is then drawn to about twice its original length, in which process the weld is recrystallized and cannot be detected with the naked eye. Such tubing has the same high reliability as the seamless variety, i.e., a porosity can be expected approximately once in $\sim$ 200 ft of tube (see Appendix for suppliers).

Figure 7.1 provides a guide to the choice of commonly-available internal diameters of tubing for helium, hydrogen, and nitrogen at room temperature and at pressures likely to be encountered in practice on the basis of a pressure drop of less than 1% per metre length. At low temperatures it can be taken as a rough approximation that pressure drop is proportional to the ratio of the absolute temperatures.

7.5 VALVES

Room-temperature valves for high and low pressures are available from a wide variety of sources (some are listed in the Appendix) and only suggestions on desirable features are appropriate here. For pressures below atmospheric down to $\sim$ 0.1 mm Hg and above atmospheric up to several atmospheres according to the bore there is probably no better and cheaper valve than that based on a domed rubber diaphragm which is screwed down on to a weir, e.g., the Saunders valve.[Sa] It is important when the larger sizes of these valves are used for sub-atmospheric pressures to obtain reinforced diaphragms. These valves are obtainable down to ¼-in. bore, and bronze bodies are available bored to any of the standard pipe sizes so that tubing can be soft-soldered straight in. These valves will withstand cooling well below room temperature provided that no attempt is made to close them while the rubber of the diaphragm is still hard.

When high-pressure valves are being selected the following principles should be borne in mind:

Of the materials used for the valve nose or needle and the seat, one should be harder than the other so that the softer deforms slightly. The latter will be subject to wear, and whether it be the nose or the seat, it should be renewable without any necessity for demounting the valve body.

The sealing gland should embody some modern type of sealing ring rather than a packing which may have to be regularly compressed by means of the old-fashioned gland nut.

Experience shows that valves in which the nose or needle does not rotate wear better than those in which it does.

Provided that one is satisfied that there will be no need to demount a valve body for a considerable time, it is as well to avoid the use of screwed couplings to the valve, since these have a way of working loose and leaking after many pressure cycles.

The hand wheel on a valve should be sufficiently small that even

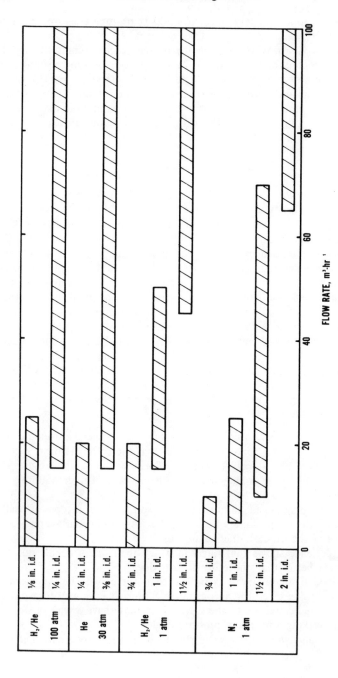

Fig. 7.1. Guide to choice of tube sizes.

the strongest user cannot damage the valve by overtightening. (It should be an invariable rule that a leaking valve be repaired: strong-arm measures always do more harm than good.)

The same principles apply to *low-temperature valves* except that lengths of poorly thermally-conducting tubing have to be incorporated between the room-temperature operating part and the valve body and nose at low temperature. The seal can almost always be situated at room temperature on the same dead-space principle as forms the basis of the Johnston coupling described in Chapter 2. (If a seal such as a bellows is used at low temperature, trouble may arise from condensation on it of water, air, etc.) A simple form of ½-in. liquid-nitrogen valve used for many years in the Clarendon Laboratory is illustrated in Fig. 7.2. Figure 7.3 shows the top and bottom of a high-pressure, low-temperature valve such as might form the Joule–Thomson expansion valve in a liquefier or refrigerator. The overall length of such a valve may exceed 1 m. The

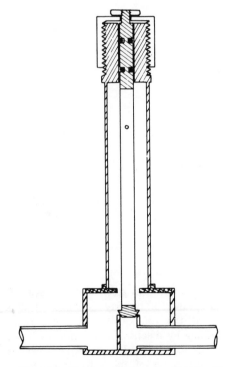

Fig. 7.2. Hand-operated valve for liquid
nitrogen.

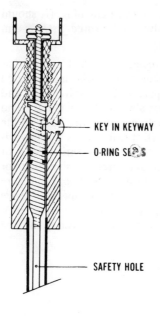

KEY IN KEYWAY

O-RING SEALS

SAFETY HOLE

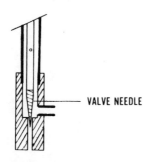

VALVE NEEDLE

Fig. 7.3. Typical valve design for gases at high pressure and low temperature.

needle may be made of stainless steel, and a semi-angle of 6° has been found suitable. The valve body may be made of nickel or phosphor bronze. The gland at the top consists of two O-rings. The needle is prevented from rotating by means of a key moving in a slot. A coarse or a multistart thread should be used so that the "feel" is not lost and so that the full operating range of the valve is traversed with not much more than one revolution of the knob.

7.6. JOINTING TECHNIQUES – METAL TO METAL

When deciding upon a method for making a joint between two metals which has to stand up to low-temperature conditions one must take the following points into consideration: the pressure across the joint; the relative thermal contraction; whether the joint is to be considered permanent, capable of being broken and remade without undue difficulty, or readily demountable; and whether the joint is so close to other joints that there is a possibility of disturbing the latter while making the former.

Inert-gas welding equipment has recently become such that the skill required to use it and the capital cost bring this technique well within the reach of most large laboratories. In Great Britain the gas used is usually argon and the term *argon-arc* welding has become common; in the USA helium is used and the term is *heli-arc* welding. The technique is applicable to many stainless steels, nickel and aluminium alloys, aluminium, and, given suitable equipment, to copper. The principle is that an arc is struck which is surrounded by an atmosphere of the inert gas, and the weld made with the aid of a suitable filler rod – a combination of the conventional arc- and gas-welding techniques. The suitability of the many stainless steels and aluminium alloys for inert-gas welding and the design of joints is beyond the scope of this book.

The stainless steels commonly used in cryogenics, namely EN58B and EN58F (AISI 321 and 347), are suitable for inert-gas welding without subsequent heat treatment. Joints made in this way are of course permanent, and, being homogeneous, are immune from stresses at low temperatures due to differential contraction.

The techniques which follow can be used, subject to certain limitations, to join different metals. In all cases it is essential to design the joint with an eye to differential contraction. The metal having the greater contraction should always be so disposed that it shrinks on to that having the lesser, so that the effect is to compress the jointing substance rather than to provoke a crack.

Silver-soldering or *hard-soldering* is a technique applicable to stainless steels, nickel alloys, bronzes, brasses, and copper which gives a strong joint which should be considered permanent, although in some cases joints can be broken and remade satisfactorily. Care needs to be taken not to overheat the joint; otherwise porosity of the solder may result. Apart from this, given reasonable cleanliness and observance of the manufacturer's instructions, it is difficult to go wrong. The temperatures required for hard-soldering will anneal hard-drawn copper, and this should be taken into account when choosing the material. Further, leaded brasses (see above) may become porous. A typical silver-solder much used in Great Britain is *Easy-Flo* manufactured by Johnson Matthey Metals Ltd.[J04] The composition specified by the makers is 50% Ag with Cu, Zn, and Cd. The alloy is nearly eutectic, having a melting range of 620–630°C. This has the advantage that there is a minimal tendency for the alloy to change its constitution during application or for it to be used in the pasty condition on account of underheating. It has a low viscosity and flows readily in the joint gap, given an adequate standard of initial cleanliness. It is important that the gap should not be wider than about 0.002 in.; otherwise surface tension will not maintain the film of solder and it will run out of the joint. When dissimilar metals are being joined it should be remembered that it is at 600°C that the gap should be of approximately this size. A smaller gap is likely to inhibit the flow of the solder. For most applications the standard *Easy-Flo* flux is suitable, but for stainless steel there is a special grade.

Soft-soldered joints are much used in the construction of cryogenic apparatus for two main reasons: they can be made close to other joints without disturbance, and they can be readily undone and remade as often as necessary. The latter point is worth bearing in mind even in the case of an apparatus which in routine use stays in one piece: replacing a piece of leaky stainless steel tubing can be almost impossible in a complex apparatus which has been assembled with hard solder or by welding.

One prejudice against soft solder can be dismissed at once. It is well known that there is an allotropic change in tin which converts the metallic "white" tin to the amorphous "grey" tin. This change occurs at a maximum rate at -48°C. Below this temperature the reaction rate becomes progressively reduced; above, at temperatures usual in temperate climates, the transformation in the absence of a catalyst is so slow as to be negligible, and does not occur at all above 18°C. This phenomenon is no bar to the use of tin/lead solders at low temperatures, because it does not occur when there is more than 5% of lead present.

The choice of alloy is crucial. The qualities required are:

1. That the melting point should be reasonably low, or the advantage mentioned above is lost.

2. That it should "wet" as many of the commonly used metals as possible without the necessity for highly corrosive fluxes.

3. That it should be a eutectic alloy, i.e., it should have a melting point at which the whole alloy becomes liquid rather than that there should be a melting range within which the alloy is pasty. The importance of this is that it is virtually impossible for a eutectic alloy – given the conditions described below – not to fill all parts of a joint with solder rather than to leave inclusions of flux or channels

which may lead to leaks. Further, the strength of a joint depends upon the alloy's setting in small crystals; this is more likely to occur in a eutectic alloy, especially if it is caused to cool rapidly.

4. That it should show no tendency to crack on cooling.

The tin/lead eutectic (Sn 62, Pb 38), which melts at 183°C, fulfils all these requirements and has proved to be easy to use and reliable over many years. It is available from such specialist firms as Enthoven Solders Ltd.(En2) or can be readily made up in the laboratory from virgin tin and lead, and cast into sticks by pouring it into an inclined length of mild steel angle. (There is a British Standard Specification 219/1959 Grade A, which is a close approximation to the eutectic and has the composition 64–65 Sn, 35–36 Pb, and less than 0.6 Sb.) It is important to take care in the purchase of tin/lead solders, because many commercial grades contain sufficient antimony to inhibit free wetting of copper and brass. The purpose of this is to produce an alloy which contains less tin and is therefore cheaper but which still melts at a temperature below 200°C. Such solders not only have the disadvantage just mentioned, but are also not eutectic alloys and have a melting range over which the alloy is pasty. However, the presence of the antimony does increase the tensile strength, although this is not an advantage, since there is no difficulty in so designing joints that the tin/lead eutectic is satisfactory in this respect.

A soft-soldered joint should be designed on the following principles (see Fig. 7.4). As a general rule, the joint should not be less than ¾ in. deep, but this naturally depends on the mechanical strength required. In the case of high-pressure work a ratio of joint length to diameter of three can be taken as a safe design figure – naturally subject to subsequent testing (see Chapter 9) – for pressures up to 150 atm. The clearance between the two surfaces to be jointed must be large enough to allow free flow of the solder down and of trapped air and flux up, but small enough for capillarity to keep the solder from running out of the joint, to reduce the possibility of cracking on

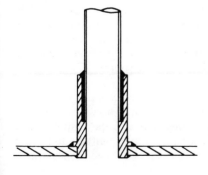

Fig. 7.4. Example of soft-soldered joint. (Only the making of the upper joint is described in the text.)

cooling, and to limit the demand placed on the shear strength of the solder. The recommended clearance is 0.003–0.004 in., but it is often possible where strength is relatively less important to get by with a larger clearance if the solder is restrained from running out of the joint as in Fig. 7.4, where the tube is shown butting up against a shoulder in the bush. The internal chamfer at the top enables solder and flux to be fed into the joint without its running down the outside of the bush.

The technique of making a soft-soldered joint is not so simple as that for a hard-soldered joint, but provided that the following routine is scrupulously observed joints of equal reliability can be guaranteed. The first step is to "tin," i.e., to coat with solder, both the surfaces to be jointed. In the case illustrated in Fig. 7.4 the tube would be tinned not only over the length of the actual joint, but for $\sim$ ¼ in. above it, and the bush on its cylindrical surface and on the surface of the chamfer. (If the clearance is such that running through of the solder is to be feared, the annular end of the tube and of the shoulder in the bush should of course not be tinned.) This tinning is most conveniently carried out by cleaning the parts with steel wool – not with emery paper, which may leave particles embedded in the metal – and then coated with a proprietary tinning paste, and warmed up, e.g., with a gas/air hand torch. (On no account should these pastes be used for making the joint, as is likely to be suggested by the manufacturers.) The tinned parts should then be washed and carefully inspected – it is easy not to notice that a narrow strip has not been coated. The two surfaces should then be lightly wetted with a zinc chloride plus ammonium chloride flux, e.g., the well-known proprietary Baker's fluid, and the parts assembled with the aid of gentle heating. Solder is then fed in until a fillet forms as shown in Fig. 7.4. Further flux may be added from a tinned copper "stick" and sufficient heat applied to keep the whole joint above, but not much above, the melting point of the solder. At this stage bubbles of air and steam will be seen to emerge through the surface, and heating should be continued until this has stopped. The solder should then be allowed to set and the joint washed outside and in with hot water. If the heating has not been carried on for long enough, dimples or even fissures will be seen on the surface of the fillet, and the joint must be wetted with flux and remade at once. It should be noted that the bush in Fig. 7.4 is so designed that the soft-soldered joint between it and the horizontal plate would not be disturbed in the course of making the joint between the bush and the tube.

Thin-walled stainless steel tubing cannot be exposed to a flame because of the near certainty of overheating. Tinning should therefore be carried out with a tinned copper soldering iron and the flame applied only to the other part of the joint during assembly. Suitable fluxes are either a mixture of equal parts of zinc chloride and concentrated hydrochloric acid, or concentrated phosphoric acid. The former is the more corrosive, but the latter suffers from the disadvantage that it becomes inactive if accidentally overheated. Very thorough washing is necessary after use, and it must be remembered that droplets of flux can be projected for surprisingly long distances away from a joint. These do little harm to thick-walled tubes, but can cause pinholes in "cryogenic" tubing.

A solder is sometimes required which does not become super-conducting at low temperatures. The eutectic alloy 60 Bi, 40 Cd, which melts at 144°C, can be used. Dilute zinc chloride in generous quantities makes a suitable flux, and great care is required to avoid overheating. This solder runs better on nickel and its alloys than on copper and its alloys.

In experimental work there is often a need for joints which can be broken and remade without disturbing nearby soft-soldered joints. Although many ingenious mechanical methods have been devised (see below), many workers still prefer to use a low-melting-point solder for reasons of simplicity of design and reliability. A favourite alloy is Wood's metal (50 Bi, 25 Pb, 12.5 Sn, 12.5 Cd), which has a melting range of 70–72°C and a tensile strength about half that of soft solder. (It is a close approximation to the eutectic, 49.5 Bi, 27.3 Pb, 13.1 Sn, 10.1 Cd, which melts at 70°C.) It is obtainable in sticks of convenient cross section from British Drug Houses Ltd.[Br3] The same flux can be used as for soft-soldering, and again care has to be taken to avoid overheating.

Stainless steel-to-aluminium joints cannot as yet be made by any soldering or welding process within the reach of the average laboratory workshop. Success has been reported with friction welding, plating followed by conventional soldering techniques and dry compression joints. However, Gilman[2] of Whittaker Corporation[Wh] reports a method of producing a joint between bulk aluminium or aluminium alloys and stainless steel from which is produced a short length of tubing of uniform bore and outside diameter. These joints are now available commercially in a wide variety of sizes and materials and seem to offer the most reliable means of joining – usually by welding – aluminium or aluminium alloy and stainless steel components for service at low temperatures to high-vacuum requirements. The technique used is to fuse two cleaned and machined surfaces by extruding them through a die at high temperature in an evacuated capsule. This results in the diffusion of one metal into the other over a region which on microscopic examination is found to be less than 1 μ thick. Solid rod is used for the process and the finished tubular joints are machined afterwards. The mechanical strength is limited only by that of the aluminium or aluminium alloy tube; rigorous thermal cycling tests have been applied, and every joint is tested for leakage on a mass-spectrometer leak detector.

Epoxy resins without additives have similar thermal expansion coefficients to the plastics listed in Table XV and their use with metals at low temperatures is limited to certain special cases which are in general of application only in experimental work. However, the addition of "fillers" decreases the thermal expansion coefficient so as to approximate to those of metals. Success is reported[3] with metal-to-metal joints at low temperatures made by means of Araldite(Ci) resins loaded with 70% w/w of powdered chalk. A hard Araldite was chosen according to the manufacturers' recommendations for the metals concerned and their instructions concerning surface treatment of the metals followed.

Demountable joints at low temperatures are often necessary. The solution which is often imposed by lack of space and which also offers high reliability is the Wood's-metal soldered joint described above. Of the many sealing methods not involving solders which have been devised, one of the most attractive is that described by Astrov and Belyanskii,[4] which works on a new principle. A PTFE washer 0.10–0.15 mm thick is clamped between two bronze flanges – so designed that the washer cannot flow – with bronze studs and nuts to compress the flanges together. Although the contraction per unit length of the washer is greater than that of the bronze when the joint is cooled, the absolute value is small enough for adequate stress to persist in the bronze studs. This joint has been shown to stay vacuum-tight down to 1.5°K; once made it will withstand tens of temperature cycles, and can be broken and remade many times without change of washer. This technique seems likely to replace those using gold, indium, and other forms of sealing ring.

A high-pressure demountable joint for such applications as the joint between the plug and the charcoal vessel of the helium purifier described in Chapter 5 can be made by wrapping PTFE tape round a tapered screw thread in exactly the same way as joints are now made in everyday plumbing practice.* Conventional cone and socket joints secured by a nut also work satisfactorily provided that all parts are made from the same material.

7.7. METAL/NON-METAL JOINTS

Metal/glass seals have been mentioned on p. 24 in connection with permanently sealed-off transfer lines, but they also have many uses in

---

* The writer is indebted to Dr. M. J. McDermott for this startling suggestion.

experimental equipment. The coefficients of thermal expansion of the commonly-used metals are four to five times larger than those of glasses such as Pyrex at room temperature, and two distinct principles have been applied to meet this difficulty in making metal/glass seals. In the long-established Housekeeper seal no attempt is made to match expansion coefficients. A sleeve of arsenic- and phosphorous-free copper is turned down to a tapering feather edge at one end. Glass is then sealed to the terminal $\sim$2 mm and differences in contraction taken up by flexure of the thin copper close to the glass surface. A glass which becomes workable at a lower temperature than Pyrex does has to be used or the copper may melt. A glass known as C8 is preferred, and this can be joined directly to Pyrex. Such seals are available from Associated Electrical Industries Ltd.(As) or through Jencons Ltd.(Je) (see Appendix for other suppliers). The other approach is to use an alloy the coefficient of expansion of which approximates to that of a glass. That most often used is a nickel alloy — ferromagnetic at low temperatures — known as *Kovar* or *Nilo-K*. This will seal to a number of special glasses of which the French glass *Soveral* is reported to give the best results. This glass can only be joined to Pyrex *via* a graded seal of two intermediate glasses such as Soveral–Monax–W1–Pyrex. (These special glasses are all available from Jencons Ltd., who also supply the ready-made seals.) Such seals are more difficult to make successfully than copper/glass seals, but have the advantage of much greater mechanical strength.

Metal/plastic seals are still being developed by individual workers, and no foolproof all-purpose technique has emerged. All that can be done is to suggest the following lines of attack. The adhesion and mechanical strength of epoxy resins is such that where the area of contact is small one can get away with unsophisticated techniques. For example, the writer found that wires could be sealed through holes in metal containers at liquid-helium temperature provided that the blob was not more than $\sim$1 mm in diameter. Methods for sealing windows at low temperatures are discussed in the next chapter.

7.8. LEAK DETECTION

As in many other experimental fields, the necessity for very low pressures frequently arises in cryogenics — usually to minimize heat influx due to thermal conduction through gases. Although working at temperatures near that of liquid helium eases the attainment of low pressures because of the low vapour pressure of all other substances once condensed on a surface at this temperature, the troubles associated with leaks are more of a trial to the low-temperature experimentalist for the following reasons:

1. Differential thermal contraction at joints – even if these are properly designed – may cause leaks to appear and, still worse, to disappear on rewarming.

2. Very small imperfections through the thickness of metal components – notably where the thickness is small – sometimes give rise to leaks only after many temperature cycles, and it seems likely that they would not appear at room temperature for a very long time.

3. It is found in practice that at the lowest temperatures the leakage rate for a gas may be two to three orders of magnitude larger than at room temperature, and still more in the case of superfluid helium.

4. Since the leakage rate is usually greater at low temperatures than at room temperature, it sometimes happens that although the presence of a leak large enough to vitiate the performance can be established when the equipment is at low temperature, it cannot for practical reasons be located except at room temperature, when the leak may be on the borderline of being detectable even with the most sophisticated instruments. Consequently, no cryogenic laboratory can afford to be without a mass-spectrometer leak detector and its use should not be confined to the testing of complete apparatus: much time can be saved if material such as thin-walled stainless steel tubing which may already have leaks in it is tested before it is built into a complicated apparatus. However, one must bear in mind that in cases of extreme difficulty in finding a leak time may be saved if locating it exactly is abandoned in favour of rebuilding the part of the apparatus in which it can be proved or is thought likely to be.

Many manufacturers of high-vacuum equipment offer mass-spectrometer leak detectors which over the years have become cheaper, more sensitive, more reliable, and almost foolproof. Points to look for, apart from obvious features such as sensitivity, convenience in operation, and protective circuitry, are the ease with which filaments can be changed and whether the instrument can be readily tuned to helium or hydrogen – or if necessary, to ^{3}He as well. The units in common use for leakage rates are: 1 litre-μ-sec^{-3} ("lusec") = 1 torr-litre-sec^{-1}×10^{-3} (= 1.3 mm^3-sec^{-1} at NTP).* A typical instrument is the Model 8 offered by Edwards High Vacuum Ltd.[Ed] which has a maximum sensitivity when used with helium of 5×10^{-12} torr-litre-sec^{-1}.

Such an instrument can be used to test a container or cryostat in the following ways:

1. With the internal part of the apparatus containing air, the vacuum insulating space is connected to the leak detector and the

* 1 torr = 1 mm Hg.

background control adjusted so that the leakage rate reads zero on the most sensitive range. The range switch is then set to the least sensitive range and the air displaced by helium gas at room temperature. If the leakage rate remains zero on the most sensitive range, the procedure can be repeated with liquid helium. (In the case of apparatus for use with liquid nitrogen it can be cooled down with the latter, emptied, and then flushed through with helium gas.) If a rigorous test is required, the apparatus should be put through several temperature cycles.

2. Should this procedure show the presence of a leak, it can of course only be located with the outer jacket removed. The inner part must first be brought up to room temperature otherwise the condensation of atmospheric water may block up the leak, as may handling of the surface with the fingers. Provided that the inner part of the apparatus can withstand atmospheric pressure externally when evacuated, it is then pumped out and connected to the leak detector. A fine jet of helium gas is then sprayed over all relevant surfaces starting at the top — allowance being made for the response time of the leak detector. In the case of recalcitrant leaks a section may be enclosed in a polythene bag which is injected with helium gas and then sealed. It can then be left for a long period shut off from the leak detector and any accumulated helium detected by opening the valve again.

3. In cases where evacuation of the leaking container is risky, as great an overpressure as possible of helium gas may be introduced and a needle-valve probe connected to the leak detector used to search for emerging helium gas. This procedure will find leaks only if they are about two orders of magnitude larger than the smallest leak detectable by the procedure last described.

Equipment involving gases at high pressure also has to be tested for leaks, although the tolerable leakage rates are very much higher. In the case of systems containing helium or neon the cost of escaping gas will be the main consideration. while in the case of hydrogen it is a matter of safety (1 cm^3-sec^{-1} of helium gas costing £0.07 per ft^3 and escaping continuously costs ~ £70 per annum at current prices). The traditional method of finding small leaks in high-pressure gas systems is by means of a soap and water lather, a paint brush, and usually also a dental mirror. This is a messy and time-consuming business, and one of the relatively cheap instruments which depend upon the detection of variations in gaseous thermal conductivity is worth having. The Edwards LT 6B, for example, can detect leakage rates of ~10^{-5} torr-litre-sec^{-1} of hydrogen or helium; the sensitivity to neon could be expected to be about three times less. For detecting leakage of hydrogen still cheaper and more portable equipment is available from such firms as Mine Safety Appliances Company Ltd.(Mi)

7.9. INSULATING MATERIALS

In Chapter 2 thermal insulation was considered in the context of storage vessels for refrigerants. As before, discussion will be limited to materials appropriate to insulation at liquid-nitrogen and at liquid-helium temperatures; interpolation for other substances is left to the reader.

Liquid-nitrogen transfer lines present a good example for which to set a criterion for thermal insulation, since the ratio of volume to surface is low in contrast with the case of a large storage vessel, e.g., that considered in Chapter 2, where it is high. Suppose that one has a pipe carrying liquid nitrogen 2 cm o.d. and carrying 50 litres-hr^{-1}, and that one is reconciled to a loss of 5% over a length of 10 m. This calls for a heat influx to the tube surface of less than 18 mW-cm^{-2}, and if unit thickness of insulation is applied, the mean apparent thermal conductivity $\bar{k}$ will have to be < 84 μW-cm^{-1}-deg C^{-1} if the outer surface is at 15°C. With this figure in mind we can assess the performance of the available insulating materials.

Powder insulation offers an appropriate performance in such a case. Certain minerals consisting mainly of silica are crushed, ground, and then heated. The result is a white powder each particle of which consists of a mass of cavities with a pore size $\sim10^{-6}$ cm. The bulk density of the powder is ~0.1 g-cm^{-3} and the particle size ~0.1 mm. When the pressure surrounding these powders is of the order of 10^{-3} mm Hg the mean free path of the gas molecules becomes large compared with the pore size, and heat flow occurs mainly by conduction through the particles and by radiation. Hunter et al.[5] give the values for $\bar{k}$ shown in Table XVIII for containing walls of two emissivities for three proprietary aerogels at pressures of 10^{-3} mm Hg. Suppliers of these and other aerogel insulating powders will be found listed in the Appendix, and references giving sources of data in the Bibliography.

TABLE XVIII

Mean Apparent Thermal Conductivities of Some Aerogel Powders

Material	$\bar{k}$ (300 $-$ 76°K) (μW-cm^{-1}-deg K^{-1})	
	$e \sim 0.02$	$e > 0.8$
Cab-o-sil(Ca)	12.1	27.1
Syloid(Si$_1$)	7.7	12.8
Microcel(Da$_2$)	6.1	9.5

The thermal conductivity of these powders can be reduced by a factor of about five by adding fine aluminium powder (see Hunter *et al.*[5]), which hinders the passage of thermal radiation, as do the reflecting layers in multiple-layer insulation (see below). However, in the course of time a phenomenon occurs for which the writer has come across neither an explanation nor a cure. The aluminium dust somehow segregates itself from the mixture and forms a conducting bridge which can cause an astonishingly large heat leak. Until a cure for this trouble is discovered aerogel insulating powders are best used on their own.

There is rarely any difficulty in maintaining a low enough pressure for these powders to give their optimum performance; where this is not the case an increase in $\bar{k}$ by a factor of about ten is to be expected if the surrounding gas is dry nitrogen at atmospheric pressure.

Solid foam materials have thermal conductivities an order of magnitude higher than those mentioned above, but may find applications in the laboratory, e.g., in the construction of small liquid-nitrogen containers where high performance is not needed. Especially useful are the "do-it-yourself" polyurethane foams, obtainable, for example, from Baxenden Chemical Co.(Ba) (see Appendix for other suppliers). Kropschot[6] gives the values shown in Table XIX for various solid foams at atmospheric pressure.

Liquid helium has, as has been made clear, a relatively very small latent heat. To take a concrete example, let us consider a cylindrical vessel 10 cm in diameter by 20 cm high which it is required to so insulate that the quantity of liquid helium evaporated other than by conduction and radiation down the neck tube is about 50 cm^3-hr^{-1}, corresponding to a heat influx of 35 mW. The allowable heat influx is $\sim$45 μW-cm^{-2}, and this clearly calls for a better form of insulation than has been described so far. We have seen in Chapter 2 that a single

TABLE XIX

Mean Apparent Thermal Conductivities of Solid Foams

Material	Density (g-cm^{-3})	$\bar{k}$ (300 − 77°K) (μW-cm^{-1}-deg K^{-1})
Expanded polystyrene	0.039	330
	0.046	260
Expanded polyurethane	0.08 − 0.14	330
Expanded rubber	0.08	360
Expanded silica	0.16	550
Glass foam	0.14	350

radiation shield at $77°K$ reduces the heat influx by a factor of ~200, but it may be inconvenient to have to cool this with liquid nitrogen in a separate container. An alternative – also mentioned before – is to use the cold of the evaporated gas to cool a radiation shield. However, insulation materials have been developed which will provide a low enough mean apparent thermal conductivity when used on their own, i.e., < 1 $\mu W\text{-cm}^{-1}\text{-deg } K^{-1}$. Such an insulation would give the required performance if the thickness were ~ 6 cm.

Insulation by multiple radiation shields was first investigated by Dewar in 1898 and subsequently applied to cryogenic technology by Petersen in 1951.[7] Ideally, a number of reflecting screens thermally insulated from each other are interposed between a warm outer surface and the cold inner surface which it is required to insulate. The reduction in thermal radiation when equilibrium has been attained is equal to $n + 1$, where n is the number of radiation shields on the assumption that their emissivities are equal to each other and to those of the boundary walls of the insulating enclosure. In practice, a compromise has to be arrived at between including as many radiation shields as possible and avoiding too much thermal conduction across them. The original form of multiple-layer insulation – generally referred to as *superinsulation* – consists of alternate layers of aluminium foil and glass fibre paper. Because these layers are in contact, thermal conduction will occur the magnitude of which will depend on how tightly the layers are packed (see Black and Glaser,[8] who show that a change of pressure from 2 to 14 lb-in.$^{-2}$ can increase thermal conductivity of superinsulation by a factor of five). As will be seen in Table XX, superinsulation of this type can have a very high performance, although it is expensive to apply because each layer has to be put on separately and fitted round, e.g., the domed ends of a cylindrically-sided vessel; a Swiss roll construction would lead to too great a thermal conductance in the helical direction.

The superinsulations giving the best performance are naturally the most expensive to apply, and in any specific case cost has to be reconciled with performance. In practice, superinsulation for use at temperatures lower than that of liquid nitrogen will usually be combined with at least one gas-cooled radiation shield, as in the case of the storage vessel described in Chapter 2. It is interesting that layers of glass fibre paper similar to that used for superinsulation give a $\bar{k}_{(300 - 76°K)}$ of only 5 $\mu W\text{-cm}^{-1}\text{-deg } K^{-1}$ in the absence of interleaving aluminium foil (Kropschot[11]).

Another approach (see Hnilicka[12]) has been to use polyester film – e.g., Melinex or Mylar – coated on one or both sides by vacuum-deposition with a film of a good reflector, usually aluminium (see Table XXI). This can be wound loosely in the flat form, or

TABLE XX

Some Representative Examples of Aluminium Foil Superinsulation

Manufacturer and code no.	Thickness of aluminium foil (mm)	Separator	Layers per cm	$\bar{k}$ ($\sim$300 to 20°K) (μW-cm^{-1}-deg K^{-1})	Ref.
Union Carbide S1-10	0.0065	Glass fibre web	4.3	1.7	9
Union Carbide S1-62	0.0065	Glass fibre paper	24	0.35	9
Union Carbide S1-92	0.0065	Glass fibre paper	46.5	0.20	9
—	0.0058	0.20 mm Glass fibre paper	22	0.39	10
—	0.013	0.13 mm Glass fibre paper	20	0.42	10
—	0.013	0.20 mm Glass fibre paper	15	0.52	10

some process of crinkling or dimpling the film or alternate layers of it can be used to provide a more readily-controlled optimal spacing. Aluminized polyester film insulations have two considerable advantages over those just described: they are much lighter in weight, and since the thermal conductance along the film is negligible, the much cheaper Swiss roll formation can be used.

However, there is one point which appears to have received less attention than it deserves. Experimental work by Ruccia and Hinckley[13] and a comment by Riede and Wang[14] indicate that the effective emissivity of a reflecting layer cannot be expected to remain as low as that of the bulk metal if the layer thickness is less than 1% of the wavelength of the thermal radiation. The values of the wavelength of maximum amplitude at three representative temperatures (White[15]) are:

T (°K)	λ_{max} (μm)
300.0	9.67
77.0	37.70
4.2	690.00

TABLE XXI

Some Representative Aluminized Polyester Film Superinsulations*

Manufacturer	Polyester film thickness (μm)	Al layer thickness (Å)	Layers per cm	Means of separation	$\bar{k}$ ($\sim$300—$\sim$77°K) (μW-cm^{-1}-deg K^{-1})	Ref.
National Research Corporation (USA)	63.5	254	?†	Crinkling	0.28	Hnilicka,[12]
—	12.7	250?	24	Loose wrapping	0.85	Scott et al.[16]
STP Ltd.(Sp)	12.7	250?	47	Loose wrapping	1.8	Kropschot[11]
Dimplar 100M42D 15	25.4	?‡ (both sides)	9.0	Alternate layers flat and dimpled	1.01	Kropschot[11] —

* It is regretted that this table is somewhat incomplete, but this is inevitable in view of the commercial interests involved.
† Bulk density 2–3 lb-ft^{-3}
‡ Measured at Clarendon Laboratory as $\sim$250 Å.

The thickness of the aluminium layer on generally-available aluminized polyester film is usually about 250 Å or 0.025 μm, and the experiments just referred to show it to have an emissivity of 0.03, compared with that for the bulk metal, which is $\leqslant 0.02$. It is therefore clear that unless attention is paid to the thickness of the reflecting layer, performance at low temperatures will be vitiated, although this effect will be counterbalanced by the fourth-power dependence on temperature of the magnitude of the thermal radiation.* The thickness of the film can be readily checked by measuring the resistance across two opposing edges of a square. In the case of an aluminium layer of 250 Å thickness, the resistance is $\sim$1 ohm.

The values of $\bar{k}$ given in Tables XX and XXI are for pressures in the region of 10^{-6} mm Hg, and in the case of vessels working at liquid-hydrogen or liquid-helium temperatures the maintenance of such pressures is no problem. Above pressures 10^{-4} to 10^{-3} mm Hg the performance of superinsulation begins to be affected by interstitial gas pressure (see Hnilicka[12]).

Note on units:
1Btu-hr^{-1}-ft^{-1}-deg F^{-1} = 17.3 mW-cm^{-1}-deg K^{-1}.
1kcal-hr^{-1}-m^{-1}-deg C^{-1} = 11.5 mW-cm^{-1}-deg K^{-1}.
10,000 Å = 1 μm = 10^{-4} cm.
1 mil (USA) = 1 thou (UK) = 10^{-3} in. = 25.4 μm.

REFERENCES

1. Hoare, F. E., Jackson, L. C., and Kurti, N. (eds.), *Experimental Cryophysics* (1961), p. 122.
2. Gilman, A. R., *Advan. Cryog. Eng.* **13**, 271 (1968).
3. Good, J. A., Clarendon Laboratory, private communication.
4. Astrov, D. N., and Belyanskii, L. B., *Cryogenics* **7**, 111 (1967).
5. Hunter, B. J., Kropschot, R. H., Schrodt, J. E., and Fulk, M. N., *Advan. Cryog. Eng.* **5**, 146 (1960).
6. Kropschot, R. H., *ASHRAE J.* **1**, 48 (1959); see also *Applied Cryogenic Engineering,* Vance and Duke (eds.).
7. Petersen, P., Swedish Technical Research Council Report No.706 (1951); *Sartryck ur TVF* **29, 4**, 151 (1958).
8. Black, I. A., and Glaser, P. E., *Advan. Cryog. Eng.* **11**, 26 (1965).
9. Private information from Union Carbide Corp.
10. Kropschot, R. H., Schrodt, J. E., Fulk, M. M., and Hunter, B. J., *Advan. Cryog. Eng.* **5**, 189 (1959).
11. Kropschot, R. H., *Cryogenics* **1**, 171 (1961).
12. Hnilicka, M. P., *Advan. Cryog. Eng.* **5**, 199 (1959).
13. Ruccia, F. E., and Hinckley, R. B., *Advan. Cryog. Eng.* **12**, 300 (1967).

*See Note 2 on p. 146.

14. Riede, P. M., and Wang, D. I.-J., *Advan. Cryog. Eng.* **5**, 209 (Q & A) (1959).
15. White, G. K., *Experimental Techniques in Low-Temperature Physics.* Oxford University Press (Clarendon Press) (1968), p. 218.
16. Scott, R. B., Denton, W. H., and Nicholls, C. M., *Technology and Uses of Liquid Hydrogen*, Pergamon, London and New York (1964), p. 116.

Notes added in Proof:

1. Recent painful experience has shown that the same applies to stainless steels: very small longitudinal voids are apparently not unusual in bar stock, but plate and sheet are almost invariably of high integrity.

2. Dr. R.A. Stradling has kindly measured the transmission of 1-mm laser radiation through a 250-Å aluminum coating on polyester film and found it to be less than 10^{-6}. This result, which is close to the calculated value, shows that for radiation characteristic of the lowest temperatures thin reflecting films are much more efficient than has been suggested in the text.

Chapter 8

Cryostat Dewars

There are two distinct ways of designing a cryostat for work at very low temperatures, which we shall take as meaning liquid-helium temperature, leaving the reader to make the appropriate simplifications for himself if he is working at higher temperatures. For straightforward experiments there may be a case for enclosing a container for liquid helium in a radiation shield and an outer vessel within which a high vacuum can be maintained. (See Croft and Thomas[1] for a small-scale cryostat built on this principle.) However, most experimental work at low temperatures is so complicated that it is usual to use a separate metal dewar vessel which can be raised up round the complete experimental assembly so that the latter is readily accessible when setting up experiments. The design requirements of such vessels are basically similar to those for the storage vessels described in Chapter 2, but low evaporation rate has to take second place to demands set by the experimental apparatus and its environment. For example, the neck tube has to be wide enough to pass over whatever is to be submerged in liquid helium and its length is usually determined by the distance between the floor and the ceiling of the laboratory – unless it happens that a pit can be made in the floor. (This point should be remembered in laboratory design; a 10-ft ceiling height can be a serious limitation.)

As with the transfer lines described in Section 2.8, there is often a strong case for making cryostat dewars in the laboratory: they can be tailored to the experiment and the production cost is likely to be very much less than the cost of those available commercially. It therefore seems worth while to describe in detail the design and construction of cryostat dewars in current use in the Clarendon Laboratory.

8.1. LABORATORY-BUILT LIQUID-HELIUM CRYOSTAT DEWARS

Figure 8.1 shows in section the generalized design of a tailed dewar for use in experiments requiring external magnetic fields. A wide

147

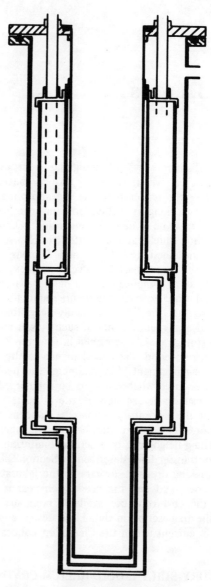

Fig. 8.1. Basic Clarendon Laboratory liquid-helium cryostat dewar.

variety of changes is rung on this design to suit different experimental requirements, and this alone proves the point that it is worth making one's own. The flange at the top is bolted to the horizontal plate through which pass the tubes, etc., leading to the experimental chamber at the bottom of the dewar and is sealed to it by means of a conventional synthetic rubber O-ring. In the design illustrated the flange would be made from sheet brass, the central bush would be hard-soldered into it, and the stainless steel neck tube soft-soldered into the bush. The advantage of this construction is that the neck tube can be replaced fairly readily. However, given adequate welding facilities, a stainless steel top flange could equally well be used and the neck tube welded to a suitably-designed upstanding edge. The outer case may be made of solid-drawn brass tubing or of rolled and seamed aluminium alloy or stainless steel. Brass tubing is somewhat more expensive and heavy, but the latter call for a higher level of skill and access to the appropriate welding equipment. In any case, the joint between the outer case and the flange is best made by means of an O-ring seal, as shown, so that the dewar can be dismantled if the need for repair arises. In the design illustrated the bottom of the wide part of the outer case is closed by a flat cap machined from sheet brass (see p. 122) which is hard-soldered to what is in this instance a brass tube.

The alternative is to use a spinning; this has the disadvantage that it increases the overall height of the dewar slightly and that where standardization is impossible extra time may have to be spent in making a former which will be used once only. If the outer case is made from aluminium alloy or from stainless steel the design must be adapted to suit inert-gas welding techniques. The liquid-nitrogen reservoir is constructed from copper and may be not less than 10 cm from the top and ~25 cm in height. It is supported by three stainless steel tubes and is thermally anchored to the central tube by an inward extension of the top plate. A copper radiation shield is soft-soldered to the bottom of the liquid-nitrogen tank. Care needs to be taken in selecting a copper with an adequately high thermal conductivity at $77°K$ and in choosing an adequate thickness if the lower part of the shield is to stay cold enough to retain the necessary efficiency. It is found in practice that so-called "high-conductivity" copper 1 mm thick is a reasonable compromise. To produce a surface of low emissivity, the radiation shield should be chemically cleaned, e.g., in chromic acid, not too long before the dewar is assembled and evacuated. The inner surface of the outer case should not be forgotten; there may be a case for copper-plating it. The extension of the radiation shield into the tail section presents a more serious problem if space is limited, as it usually is. A satisfactory solution is to form a tube of the right size by electrolysis. (This is a specialized technique and reference should be made to one of the standard texts on the subject.) Communication has to be provided for gas to pass through the radiation shield during pumping: this is the purpose of the holes shown at the bottom of the wide part. The clearances in the tail section can be kept to a minimum by using 0.2-mm-diam nylon monofilament wound in spirals as spacers; the effect on the evaporation from the dewar has proved undetectable. The liquid-helium container is made from brass tube turned down to a wall thickness of 1 mm with end caps turned from sheet brass as before – all joints being hard-soldered. The tail section is made from the Rollet brass tubing mentioned on p. 126. A pumpout valve and overpressure relief valve are fitted to the outer vacuum jacket. The evaporation from such a dewar used with an experimental assembly of average complexity is about 100 cm^3-hr^{-1} of liquid helium.

Where tail sections for insertion into external magnet solenoids are not required the alternative of superinsulation can be used. An example of a simple dewar vessel of this sort has recently been made in the Clarendon Laboratory (Good*) to contain a superconducting solenoid. A stainless steel tube 12.5 cm in diameter by 0.3-mm wall thickness and 76 cm long forms both the neck tube and the liquid-helium vessel. The bottom half of this tube is surrounded by a copper radiation shield thermally bonded at its top to the midpoint of the stainless steel tube, thus forming a gas-cooled radiation shield. This assembly is wrapped with eighteen pairs of *Dimplar* multiple-layer insulation (see p. 144) in Swiss roll fashion making a layer about 1 in. thick. Fifteen pairs of Dimplar disks insulate the bottom. The Dimplar is spaced away from the radiation shield by means of adhesive polyurethane foam strip of the sort used for domestic draught exclusion. The outer container is made from welded aluminium sheet. A brass plate is hard-soldered to the stainless steel tube and a vacuum-tight joint made to the aluminium container by means of a conventional O-ring seal. The evaporation from the dewar vessel alone with six radiation shields in the neck was less than 100 cm^3-hr^{-1} when the depth of liquid helium was 30 cm. The temperature of the radiation shield varies from 40° to 80°K, depending on the level of the liquid helium. Provided that a simple configuration such as this can be used, a liquid-helium cryostat can thus be built by means of relatively straightforward techniques.

Where liquid-helium reservoirs are made from metals having relatively poor electrical conductivities and therefore relatively high emissivities (see Table XV) a reduction in the heat flux from the innermost radiation shield may be achieved by electroplating the former with a layer of, for example, copper. If this is to be effective, it should not be much less than $\sim 1\,\mu$m thick (see p. 143), and should of course be kept chemically clean.

Some experiments require cryostat dewars with windows able to transmit X-rays, ultraviolet or visible light, infrared radiation, or particle

---

* The writer is indebted to Dr. J. A. Good for providing this information in advance of publication.

Fig. 8.2. Philips liquid-helium cryostat dewar with tail
section demounted. (Photo: Philips, Eindhoven.)

beams. The construction of suitable room-temperature windows is a
matter of standard technique, but sealing a quartz window, for example,
into a metal wall so that it is vacuum-tight at low temperatures presents
problems arising from differences in thermal expansion. These have to
be considered in relation not only to that between the window and the
metal part of the apparatus, but also between the components and the
adhesives used to stick them together. No ideal solution can be offered:
it is more a matter of describing what ingenious experimenters have
managed to get away with. Roberts[3] has described a method applicable
to most rigid window materials which is based on the same principle as
the Housekeeper copper/glass seal. A diaphragm of annealed copper
0.003 in. thick is formed into the contour of the frustum of a cone of
large angle. The window is sealed into the smaller aperture with Araldite
Type 1 and the larger end attached to the main body of the cryostat
with Wood's metal or some other low-melting-point solder. Vos and
Kingma[4] report success with a quartz window sealed to a polished
Monel face by means of an indium ring 0.8 mm thick by 0.5 mm wide
in cross section, smeared with silicone tap-grease. Helium gas at room
temperature and a pressure of not more than 20 atm was applied so as
to cause the grease to permeate the sealing surfaces. Where polyester

film is suitable, e.g., in X-ray work, a thin film of an epoxy resin of low viscosity can be used to stick the film to a metal surface. (See, for example, Thomas,[5] who reports success with Araldite type MY753 used with HY951 hardener.)

8.2. COMMERCIAL LIQUID-HELIUM CRYOSTAT DEWARS

Many firms offer a multiplicity of types of liquid-helium cryostat dewar, and some will build complete cryostats to order (see Appendix). Several manufacturers offer standard liquid-helium containers with inter-changeable tail sections to suit different types of experiment. Figure 8.2 shows a Philips cryostat dewar with the three parts of a plain tail

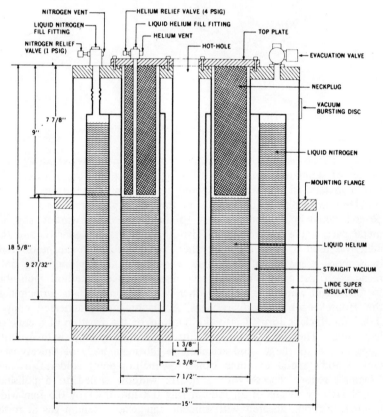

Fig. 8.3. "Hot-Hole" liquid-helium dewar for superconducting solenoid by Linde Division of Union Carbide Corp. (Drawing: Union Carbide Corp.)

section. It holds 1.25 litres of liquid helium which with a charge of liquid nitrogen will last for at least 24 hr. Figure 8.3 shows a super-insulated liquid-helium dewar made by the Linde Division of Union Carbide to hold a superconducting solenoid with an o.d. of less than $7\frac{1}{2}$ in. and an i.d. of greater than $2\frac{3}{8}$ in. A so-called "hot hole" of $1\frac{3}{8}$ in. diameter is provided through the centre of the vessel. As the physical size of superconducting solenoids increases, larger vessels are becoming available to contain them.

It is not unknown for commercial cryostat dewars to leak, and the prospective purchaser should ask what the testing routine is and whether he can expect immediate replacement should the vessel fail within the guarantee period.

REFERENCES

1. Croft, A. J., and Thomas, J. O., *Cryogenics* 9, 57 (1969).
2. Roberts, V., *J. Sci. Instr.* 31, 251 (1954).
3. Vos, J. E., and Kingma, R., *Cryogenics* 7, 50 (1967).
4. Thomas, J. O., *Acta Crystallographica*, in preparation.

Chapter 9

Hazards

9.1. PHYSICAL HAZARDS

As has been made clear in earlier chapters, liquefiers and refrigerators and their associated gas-storage systems include many components containing gases at pressures above atmospheric. The energy stored in a compressed gas may be considerable − in contrast with the case of a compressed liquid − and mechanical failure may have such serious consequences that the following steps should be followed scrupulously, however strong the temptation to take short cuts on account of lack of time, access to test equipment, etc.

Bought-out components can be trusted to be safe up to the stated working pressure, although in the case of items to be used at low temperature it is up to the user to satisfy himself about the suitability of the material − see especially the *caveat* about carbon steels on p.123. A criterion has been given in Chapter 7 for the selection of tubing, namely, that the working pressure should not be more than a fifth of the bursting pressure. However, it is not enough to choose appropriate components, materials, and methods of assembly and merely to hope for the best. It must be an invariable rule that a complete assembly be tested hydraulically. The writer's practice is to test to twice the working pressure, but this may err on the side of overcaution, and a factor of 1.5 may be more in keeping with engineering practice. A hand-operated hydraulic test pump working on water is not an expensive item and one that no cryogenic laboratory can afford to be without (see Appendix for makers).

A high-pressure system for a gas if properly designed, made, and tested may be entirely safe in itself, but usually it is the case that it communicates with a low pressure system. If this is so, one must look for all possible ways in which the pressure in the latter may become too high to be safe − e.g., the failure of a high-pressure valve to shut off properly may be combined with the mistaken closing of low-pressure valves, with the result that an unsafe pressure could build up in a closed volume. Wherever such abnormal conditions might arise a relief valve

154

must be fitted. These are available from most of the firms selling other types of valve. Alternatively, the simple type shown in Fig. 9.1 can be made up in the laboratory. Once a high-pressure gas system has been constructed, tested, and put into commission it should not be taken for granted that it will remain safe indefinitely. Manufacturers of high-pressure storage cylinders recommend inspection and testing every five years – a job for them and not for the user. This time can probably be safely exceeded where cylinders are static and used for noncorrosive gases such as hydrogen and helium, and if care is taken to protect them from rusting externally. A careful eye should be kept on the possibility of fracture of copper pipes where these have become subjected to work-hardening, although in a well-designed system this will not happen. There are reports that copper pipes have failed as a result of over-enthusiastic polishing.

Portable high-pressure gas-storage cylinders can be a source of hazard if they are not handled carefully. This is now well enough known for delivery men not to drop them on to a concrete surface. However, serious accidents are possible if a free-standing cylinder is knocked over and the valve happens to hit some solid object. For this reason cylinders should always be chained up and not left standing where they could be knocked over – even if they are empty. They should be transported on wheeled trollies (a particularly neat and simple type is manufactured by Scientific Supplies Co.(Sc) on to which a cylinder can be put without its having to be lifted and which has no tendency to run away or over-balance).

Where parts of a system containing gas are at low temperature there is the possibility of pressure rise resulting from increase of temperature under constant-volume conditions. The possibility of this has already been mentioned in connection with the helium gas purifier described at the end of Chapter 5, where the necessity for the inclusion

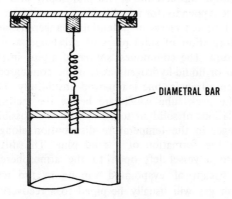

Fig. 9.1. Low-pressure relief valve.

of a pressure-relief valve was stressed. If the liquid nitrogen were to be allowed to run out with a high pressure of gas in the bottle and with the inlet and outlet valves both closed, the pressure would rise when the bottle had warmed up to room temperature by a factor of about four, and this could lead to disastrous consequences. Although two simultaneous acts of negligence are required to bring about such a state of affairs, the possibility must nevertheless be guarded against.

In the case of liquid-refrigerant containers the theoretically possible pressure which could result if a full container were allowed to warm up to room temperature at constant volume is of course very much higher – e.g., greater than 1000 atm for liquid helium. It is therefore necessary to give considerable thought to the design and use of all containers for the storage of cryogenic refrigerants. The most obvious hazard is a blockage of the neck tube. In due course, unavoidable heat influx will cause the consequent rise of pressure to rupture the inner vessel containing the refrigerant. This pressure will then get into the vacuum enclosure and the pressure in it will be greatly increased by the rapid boiling of the refrigerant. The outer case must therefore have a strategically sited area of reduced mechanical strength such that the pressure can be released so as to cause the least possible risk of injury or damage. When buying refrigerant containers one should look for the inclusion of such a feature even though it is unlikely to be required if the precautions to be described are followed.

It is not beyond the bounds of possibility that some misguided person may push a bung into the opening of a refrigerant vessel so efficiently that the pressure required to rupture the vessel is smaller than that required to blow out the bung. A relief valve must therefore be fitted below the opening such that the safe working pressure of the vessel cannot be exceeded. Experience has shown that this alone is not enough: the relief valve must be such that it is not possible to push a bung into its outlet or otherwise to defeat it. In some cases it may be advisable to adopt the belt-and-braces philosophy and to fit a bursting disc as well (see Appendix for suppliers).

A more frequent cause of potentially dangerous blockages of neck tubes is the formation of solid plugs of substances which are fluids at room temperature. The commonest source of a plug in the neck tube of a liquid-helium or liquid-hydrogen vessel is the condensed air which may drip down when a transfer line is removed unskillfully. This may solidify on parts of the neck tube which are below the melting point. Subsequent accumulations of solid air together with the possible concentrating effect of changes in the temperature distribution along the neck tube may result in the formation of a solid plug. The diffusion of atmospheric air into a vessel left open to the atmosphere can occur, in spite of the stream of evaporated vapour in the reverse direction; however helium gas will usually be piped to a recovery system. Where this is not possible, e.g., when vessels are in transit, some form of relief·valve set to a suitably low pressure should be fitted. A simple form is the Bunsen valve: a longitudinal slit in a piece of rubber tubing sealed at one end. From such causes plugs in the neck tubes of liquid-refrigerant containers do occur from time to time even in the best-regulated laboratories. The following means of clearing them should be generally known and the necessary equipment kept handy. The technique is to blow a stream of room-temperature helium gas on to the plug. It may happen that a transfer line cannot be withdrawn because of the plug, and then a tube of small diameter is needed to get down the space between the outside of the transfer line and the inside of the neck. This treatment is very effective, but needs carrying out promptly: one should have no hesitation in using a hacksaw if such violent means are the only way of enabling one to insert the tube.

Some manufacturers of refrigerant-storage vessels have reduced the possibility of dangerous conditions arising from plugged neck tubes by fitting a wide substantive neck tube with a coaxial tube inside it large enough to take transfer lines. A room-temperature pressure-relief valve is fitted so as to communicate with the annular space between the two tubes. The inner tube can usually be removed easily and plugs cleared by merely letting them warm up. Although this arrangement contributes to safety, the fact that the evaporating helium does not take up some of the heat passing down the neck tube increases the loss rate from the vessel.

There are other circumstances in which an unexpected and possibly hazardous overpressure can develop. An actual case is illustrated in Fig. 9.2, where are shown the remains of one of the earlier generation of liquid-helium storage vessels in which a 17-litre copper dewar vessel is suspended inside a concentric copper liquid-nitrogen vessel. The container had come back from being out on loan with its inner vessel

Fig. 9.2. Remains of exploded 17-litre liquid-helium vessel. (Photo: C. W. Band, Clarendon Laboratory.)

empty but probably a good deal of liquid nitrogen in the outer vessel. It stood unused for a fortnight and was found one morning in more or less the condition shown. Minor damage to the ceiling timbers showed that part of the vessel had hit the ceiling with considerable force. It will be seen that the three outer spheres have been burst open at their equatorial seams, while the inner sphere has been crushed. The sequence of events can be deduced as follows: while the inner vessel was cold a small leak must have developed from the liquid-nitrogen space into the insulating vacuum space round the inner sphere. The cold outer surface of the inner sphere would have maintained the vacuum until the liquid inside had boiled away when the pressure would have risen to a value which could not be higher than atmospheric. This state of affairs would have persisted until the liquid nitrogen in the outer vessel had evaporated and the inner vessel was beginning to warm up. The size of the leak must have been such that the gas in what should have been the vacuum space of the inner vessel was unable to get out fast enough to prevent the collapsing of the inner shell and the explosive rupture of the outer vessel must have followed as a consequence. It is difficult to see what steps could have been taken in the design of these vessels to anticipate this improbable sequence of events. However, in the modern superinsulated vessel there is only one vacuum space, and a bursting point at room temperature is easily provided. The moral of this story is that designers of cryogenic equipment should allow their imagination free reign in foreseeing the consequences of all possible leaks and blockages and take appropriate steps.

9.2. CHEMICAL HAZARDS

Liquid oxygen is rarely used as a refrigerant nowadays, but where there is an advantage in price or availability it should not be rejected out of hand for safety reasons. Care needs to be taken to avoid concentrations of vapour which might cause a fire to start up which otherwise would not. Two definite cases for not using liquid oxygen are where hydrogen is present, or where it is proposed to use an oil-sealed pump, since there may be an explosive reaction with the oil. This also applies to liquid air, in which the oxygen content increases as it evaporates.

The economic advantage of liquid hydrogen as a refrigerant — notably for precooling cryostats before they are filled with liquid helium — has been explained in Chapter 1. Although there is no denying that hydrogen is a dangerous substance to have about, the distinction between the hazardous nature of a vessel containing liquid hydrogen and that of a cylinder of hydrogen gas at room temperature is sometimes exaggerated. The hazard from the slow evaporation of hydrogen gas from the neck of the former or from the more rapid evaporation when a transfer tube is inserted can be minimized by the use of suitable arrangements for leading the hydrogen away to the atmosphere outside the laboratory. The remaining danger is that should some serious accident befall the vessel, a considerable quantity of hydrogen will be released in a short time and there will be nothing akin to a cylinder valve by which to stop it. If one can be confident of being able to maintain the same control over the escape of hydrogen into the atmosphere from a cryogenic system involving liquid hydrogen as from a system taking hydrogen gas from a cylinder, then the problem of handling liquid hydrogen is reduced to the same proportions. This confidence can only follow from high standards of care in the use of appropriately chosen and maintained equipment.

There are bound to be occasions when some hydrogen gas must be allowed to escape into the atmosphere of a laboratory at certain definite times and points, although with careful technique the quantity need only be very small. Where this occurs there will inevitably be an inflammable mixture over a small region. (More than 4% by volume of hydrogen in air constitutes an ignitable mixture.) Care therefore needs to be taken to avoid possible sources of ignition at these points. Assuming that no one is going to be foolish enough to use ordinary electrical apparatus, strike matches, etc., where explosive concentrations are likely to occur, we can pass to two recondite means by which sparks may be generated. One is by the discharge of an electrostatically-charged person happening to occur in a region where there is an inflammable mixture. Working in Oxford, wearing leather-soled shoes and cotton and wool clothing, the writer has had no direct experience of this phenomenon; those living in dryer climates, wearing rubber-soled shoes and clothing made from synthetic fibres frequently find themselves charged up, and this can be a very real danger. There are also certain floor finishes which make matters worse. A simple instrument can be made up which enables one to test one's resistance to earth, and if this is less than 2 megohms one has nothing to worry about. Provided that floor surfaces are sufficiently good conductors, personal electrostatic charging can be avoided by wearing conducting overshoes. Electrostatic charging of equipment can also occur, as when an earthed metal dewar vessel was being filled from a large storage vessel which happened to be fitted with synthetic rubber casters. A spark was seen to pass between the transfer tube and the mouth of the vessel which was being filled. The hydrogen caught fire – there was no explosion – and the flame was easily put out with a carbon dioxide extinguisher. On investigation it was found that the large storage vessel was insulated from earth to better than 100 megohms. Trouble of this sort can be avoided by means of earthing clips.

If regions where inflammable concentrations of hydrogen in air occur are localized and temporary and if sources of ignition in these regions can be obviated, then liquid hydrogen can be used safely in the laboratory without other special precautions. (This is on the assumption that glass cryostat dewars containing liquid hydrogen are a thing of the past: spontaneous shattering was not uncommon. Spherical glass transport dewars up to 5-litre capacity are safe if well made and carefully handled.)

Where concentrations of hydrogen above 4% are likely to occur over the whole of an enclosed space, an entirely different approach has to be made. Where electrical apparatus cannot be avoided it must follow the lines laid down by an accepted standard such as the British Standards Institute Code of Practice number 1003, which recommends the relatively-cheap pressurized electrical apparatus. The philosophy here is that where there are leaks in the sealing of the appliance there will be an escape of the inert pressurizing gas rather than entry of the possibly inflammable surrounding atmosphere. This leads to much less bulky and expensive appliances than the earlier "flameproof" type, which were based on the principle that an explosion inside the appliance could not start one outside. The standards covering these fittings do not in any case cover them for use in atmospheres containing hydrogen, but some flameproof electric heaters can be economically applied in pressurized systems. Electrical equipment handling small currents and voltages which conforms to BSS 1259 is termed "intrinsically safe," i.e., the user can be satisfied that it will not ignite an inflammable gas mixture inside or around it.

9.3. PHYSIOLOGICAL HAZARDS

Apart from injuries caused by explosions due to the types of situation already discussed, and given that no one is going to be stupid enough to submerge a finger in liquid nitrogen to see what happens, there are few sources of personal hazard in cryogenic work. Where the once ubiquitous glass dewar vessels are still in use the possibility of damage to the eyes by flying glass fragments should be guarded against. An ophthalmic surgeon has given the opinion that those normally wearing spectacles are not at risk and that for others a pair of safety spectacles without shields at the sides are adequate. Serious burns from cold objects or refrigerants are rare. Minor splashes of liquid nitrogen do no harm unless they happen to lie in the so-called anatomical snuff box on the back of the hand near to the thumb. A more insidious danger is the possibility of the depletion of the oxygen concentration in the atmosphere – notably by excess nitrogen. A typical precaution against this is illustrated in Fig. 5.4, where a fan grille can be seen through which the evaporated nitrogen is sucked away when the helium purifiers are first cooled down. A lack of oxygen may also occur when the pressure is released from a liquid-nitrogen storage vessel.

For an exhaustive treatment of hazards – though more from an industrial than from a laboratory point of view – see Zabetakis.[1]

REFERENCE

1. Zabetakis, M. G., *Safety with Cryogenic Fluids,* Plenum Press, New York, (1967).

Bibliography

GENERAL (in order of usefulness as a single up-to-date companion volume to this book.)

White, G. K., *Experimental Techniques in Low-Temperature Physics*, Clarendon Press, Oxford, Second Edition, 1968.
Physical principles and practical details of low-temperature experimental techniques, with useful data and sources of supply of equipment and materials. Little overlap with this book.

Hoare, F. E., Jackson, L. C., and Kurti, N., eds., *Experimental Cryophysics*, Butterworths, London, 1961.
19 contributors cover a wide field of experimental technique, but inevitably the treatment of some topics is now somewhat out of date. Contains useful data.

Rose-Innes, A. C., *Low Temperature Techniques*, English Universities Press, 1964.
A practical guide to the use of liquid helium and to the design and use of cryostats, including temperature measurement and the use of liquid helium-3. Includes useful data and sources of supply of equipment and materials.

Scott, R. B., *Cryogenic Engineering*, Van Nostrand, Princeton, 1959.
An essentially practical book covering large-scale plant as well as laboratory-scale equipment, but not experimental technique. Authoritative but now inevitably somewhat out of date in places.

Vance, R. W., and Duke, W. M., eds., *Applied Cryogenic Engineering*, Wiley, New York 1962, and Vance, R. W., ed., *Cryogenic Technology*, Wiley, New York, 1963.
These two books are written primarily for the space technologist but contain some useful chapters and data for the laboratory worker.

Timmerhaus, K. D., ed., *Advances in Cryogenic Engineering*, Plenum, New York, Vol. 1, 1960 et seq. (annually).

Mendelssohn, K., ed., *Progress in Cryogenics*, Heywood, London, Vol. 1, 1959, Vol. 2, 1960, Vol. 3, 1961, Vol. 4, 1964.

PHYSICAL DATA
Brookhaven Bubble Chamber Group Selected Cryogenic Data Notebook, (1966).

Cryogenic Materials Data Handbook ML-TDR-64-280, Martin Company under Air Force Contract AF33 (657)-9161 (1964).

Corruccini, R. J., and Gniewek, J. J., *Specific Heat and Enthalpies of Technical Solids at Low Temperatures*, Natl. Bur. Stds. Monograph 21 (1960).

Corruccini, R. J., and Gniewek, J. J., *Thermal Expansion of Technical Solids at Low Temperatures*, Natl. Bur. Stds. Monograph 29 (1961).

Gopal, E. S. R., *Specific Heats at Low Temperatures*, Plenum Press, New York (1966).

Mann, D. B., *Thermodynamic Properties of Helium*, Natl. Bur. Stds. Technical Note 154 (1962).

Woolley, H. W., Scott, R. B., and Brickwedde, F. G., *Thermal Properties of Hydrogen*, Natl. Bur. Stds. Research Paper RP 1932, Vol. 41, p. 379 (1948).

McCarty, R. D., and Stewart, R. B., *Tables of Thermodynamic Properties for Neon*, Natl. Bur. Stds. Report 8726 (1965).

Johnson, V. J., ed., *A Compendium of the Properties of Materials at Low Temperature*, 4 vols., WADD Technical Report 60-56, Pub. Nat. Bur. Stds. (1960-1961).
Densities, thermal expansion coefficients, thermal conductivities, specific heats, enthalpies, transition heats, phase equilibria, dielectric constants, adsorptions, surface tensions, and viscosities of fluids; thermal expansion coefficients, thermal conductivities, specific heats, and enthalpies of solids; compressibilities of fluids, thermal conductivity integrals of solids, entropy of neon, velocity of sound in fluids, equilibrium concentrations of binary mixtures of fluids, and electrical resistivities of metallic elements; bibliography.

Teed, P. L., *The Properties of Metallic Materials at Low Temperatures*, Wiley, New York (1950)

Touloukian, Y. S. ed., *TPRC Series on Thermo-physical Properties of Matter*, 13 vols., Plenum Press, New York, 1970.
This compilation will include much information of no relevance to cryogenics but is likely to prove the definitive source of thermal conductivity, specific heat, thermal radiative properties, thermal diffusivity, viscosity, and thermal expansion.

CHAPTER 1

Davies, Mansel, *The Physical Principles of Gas Liquefaction and Low Temperature Rectification*, Longmans, 1949.

Ruhemann, M., *The Separation of Gases*, Clarendon Press, Oxford, 1940.

Scott, R. B., Denton, W. H., and Nicholls, C. M., eds., *Technology and Uses of Liquid Hydrogen*, Pergamon, 1964.

International Institute of Refrigeration, 1965 conference at Grenoble, *Liquid Hydrogen*, Pergamon, 1966.

This book is mainly concerned with large-scale applications, but it includes some useful data and references to sources of other data. There are chapters concerned with application to nuclear physics and electrical engineering.

Farkas, A., *Orthohydrogen, Parahydrogen and Heavy Hydrogen*, Cambridge University Press, 1935.

Kropschot R. H., Birmingham B. W., and Mann, D. B., eds., *Technology of Liquid Helium*, National Bureau of Standards Monograph III, US Dept. of Commerce, 1968.

This book, to which leading authorities have contributed, includes chapters on properties of helium, liquefaction, refrigeration, storage and handling, cryoelectronics and applications to superconducting and resistive magnet technology, bubble chambers etc.

International Institute of Refrigeration, Proceedings of 1965 conference at Boulder, Colorado, *Liquid Helium Technology*, Pergamon, 1966.

A collection of articles by world authorities – some general and some specific – covering liquid-helium technology at CEL, Boulder, physical properties of liquid helium-4 and liquid helium-3 of technological interest, various aspects of heat transfer, control of temperature in cryostats, calibration of germanium thermometers, thermoacoustic oscillations, helium liquefiers and refrigerators, large-scale distribution of liquid helium, a 670-litre light-weight liquid helium vessel, applications to nuclear physics and space technology, and devices exploiting superconductivity.

Keesom, W. H., *Helium*, Elsevier, 1942.

CHAPTER 3

Collins, S. C., and Cannaday, R. L., *Expansion Machines for Low Temperature Processes*, Oxford University Press, 1958.

CHAPTER 4

White, F. G., *Industrial Air Compressors*, Foulis, London, 1967.

Yarwood, J., *High Vacuum Technique*, Chapman & Hall, 4th edition, 1967.

CHAPTER 6

Young, A. J., *Process Control*, Instruments Publishing Co., Pittsburgh, 1957.

CHAPTER 7

Yarwood, J., *High Vacuum Technique*, Chapman & Hall, 4th edition, 1967.

Rosenberg, H. M., *Low Temperature Solid State Physics*, Clarendon Press, Oxford, 1965.

Theoretical treatment of and quantitative data for physical properties of solids at low temperatures.

See also under *Physical data* above.

USA government-sponsored publications are obtainable in Great Britain through

Universal Subscription Service Ltd.,
Universal House,
4 Footscray Road,
Eltham
London, S. E. 9.

Appendix

SOURCES OF SUPPLY

This Appendix does not attempt a wide coverage of American suppliers. Readers are referred to the annual "Data Book and Buyer's Guide" issue of *Cryogenic Engineering News,* published by Business Communications Inc., 2800 Euclid Avenue, Cleveland, Ohio 44115.

General

The following firms offer a variety of cryogenic hardware including storage vessels, transfer lines, cryostat dewars, etc.: Ai_1, Ai_2, Ai_3, An_1, Ar, Br_1, Co_2, Cr_3, Cr_5, Ct, Ga_1, Ho_1, Le, Ma_3, Ox, Ph, Sp, Su_1, Su_2, Un_1, Va, Wo.

Chapter 2

p. 14. Liquid nitrogen vessels: Ro_1, Ry
        All expanded polystyrene laboratory vessels: We
p. 15. Liquid nitrogen tanks: Bu_1
p. 16. Liquid hydrogen and liquid helium vessels: Ai_1, Ar, Br_1, Co_2,Cr_3, Ga_1, Ho_1, Sp, Su_1, Un_1, Va, Wo
p. 22. Transfer lines: Al_3, Hy_1, In_4, Ja, Ti
p. 24. Metal bellows: Av, Dr, Po_1, Te_2

Chapter 3

See text for manufacturers of liquefiers and refrigerators.
p. 69. Extended surface tubing for heat exchangers: Ca_1, Ea, Yo

167

Chapter 4

p. 83.   Rotary exhausters: Br_5, Hi_2
pp. 83–4.
          Large rotating-vane vacuum pumps: Al, Ed, Le
pp. 84–6.
          Kinney-type vacuum pumps: Ed, Ge_2, Ki
          Vapour-booster pumps and Roots pumps: Ed, Ge_2, Le
p. 86.   Compressors: An_2, Co_1, Ga_2, Hy_1, Le, Lu, Re, Wi

Chapter 5

p. 88.   Wet gas holders: De_2, Ha_3
p. 88.   Dry gas holders: Po_2, Ar
p. 89.   Gasbags, synthetic rubber: Du, Ar
p. 90.   Rubber Balloons: Ch_2, Du, Gu
p. 92.   High-pressure cylinders, etc.: Ch_1, Jo_2, Un_2.
p. 93.   Drying agents: Pe, Si_2, Un_1
          Drying units: Bi, Un_3
p. 93.   Charcoal: Br_2, Ca_4
p. 94.   Platinum catalyst: En_1

Chapter 6

p. 100.   Pressure gauges—See Table XII, page 103.   Also Sm.
p. 102.   Pressure transducers: Bo, Da_1, Fi_2, In_3, KD, Ro_4, Se_2, St
pp. 104–5.
          Semiconductor thermometers: Cr_4, CS, Ho_2, Ra
p. 106.   Vapour-pressure and gas thermometers: Cr_2, Ro_4
p. 105.   Thermocouple wire: Jo_4
p. 104   Platinum resistance thermometers: De_1
p. 104.   Resistance thermometer wire: Jo_4
p. 107.   Float-in-tube flowmeters: Fi_1, Gr_1, Pl, Ro
p. 107.   Orifice plate and Venturi tube flowmeters: Ke, Wa
p. 107.   Turbine flowmeters: Fi_1, Fo_1, Me
p. 109.   Integrating flowmeters ("gasmeters"):
                dry: Pa, Wa
                wet: Al_2, Pa, Wa
p. 110.   Optical dipsticks: Hi_1, Sp
p. 112.   Carbon resistors suitable for thermometry: Bk (UK agent for Alan
                Bradley)
p. 113.   Mass-spectrometer gas analyzers: El_2, As, Co_4, Va

Chapter 9

p. 154. Hydraulic test pumps: Ma_4, Ta
p. 155. Cylinder trolleys: Sc
p. 155. Cylinder wall clamps: Wh_2
p. 156. Bursting discs: Di, Jo_4, Ma_5

KEY TO ABBREVIATIONS

Ai_1 L'Air Liquide, 75 Quai d'Orsay, Paris 7*e*, France
Ai_2 Air Products and Chemicals Inc., Advanced Products Dept., Allentown, Penna., USA.
     UK office: 49 Poland Street, London W1, England.
Ai_3 Airtec Inc., 264 Columbus Ave, Roselle, NJ 07203, USA.
Al_1 Alley Compressors Ltd., Cathcart, Glasgow S4, Scotland.
Al_2 Alexander Wright & Co (Westminster) Ltd., 91 Wellesley Road, Croydon CR9 2SA, England.
Al_3 Almac Cryogenics Inc., 1108 26 Street, Oakland California 94607, USA.
An_1 Andonian Associates Inc., 26 Thaver Road, Waltham, Mass., USA.
An_2 Andreas Hofer Ltd., Mühlheim, W. Germany.
Ar   CTI (Cryogenic Technology Inc), Kelvin Park, 266 Second Avenue, Waltham, Mass. 02154, USA.
     European office: Arthur D. Little A. G., Zürich, Seefeldstrasse 224, Switzerland.
     UK office: Arthur D. Little Ltd., Berkeley Square House, Berkeley Square, London, W.1., England.
As   Associated Electrical Industries Ltd., Scientific Apparatus Dept., Barton Dock Road, Urmston, Manchester, England.
Av   Avica Equipment Ltd., Mark Road, Hemel Hempstead, Herts, England.
Ba   Baxenden Chemical Co., Baxenden, Near Accrington, Lancs, England.
Be   Beckman Instruments Inc., 2500 Harbor Boulevard, Fullerton, California 92634, USA.
     UK branch: Beckman Instruments Ltd., Queensway, Glenrothes, Fife, Scotland.
Bi   Birlec Ltd., Tyburn Road, Erdington, Birmingham 24, England.
Bk   B. & K. Laboratories Ltd., 4 Tilney Street, Park Lane, London, W.1., England. (UK agent for Alan Bradley)
Bl Black Automatic Controls Ltd., Leafield, Corsham, Wilts, England.

Bo   Electronics Dept., Boulton Paul Aircraft Ltd., Wolverhampton, England.

Br_1 British Oxygen Cryoproducts Ltd., Deer Park Road, London, SW19, England.

Br_2 British Carbo-Norit Union Ltd., London Road, West Thurrock, Greys, Essex, England.

Br_3 British Drug Houses Ltd., Poole, Dorset, England.

Br_4 British Ermeto Corp. Ltd., Hargrave Road, Maidenhead, Berks, England.

Br_5 Broom & Wade Ltd., PO Box No. 7, High Wycombe, England.

Bu_1 W. P. Butterfield Ltd., Shipley, Yorks, England.

Bu_2 Budenberg Gauge Co., Broadheath, Near Manchester, England.

Ca_1 Wolverine Tube Division, Calumet & Hecla, 1233 Central Street, Evanston, Illinois, USA.

Ca_2 Cambridge Industrial Instruments Ltd., Sydney Road, Muswell Hill, London, N.10, England.

Ca_3 Carpenter Steel Co. Alloy Tube Division, 101 West Bern Street, Reading, Penna., USA.

UK agent: Tube Sales (U.K.) Ltd., West Bay Road, Southampton, England.

Ca_4 The Cabot Corporation, 125 High Street, Boston 10, Mass., USA.

UK office: Cabot Carbon Ltd., Caroline House, Dingwall Road, Croydon, Surrey, England.

Ch_1 Chesterfield Tube Co. Ltd., Chesterfield, Derbyshire, England.

Ch_2 Chemring, Ltd., Alchem Works, Rodney Road, Fratton Industrial Estate, Portsmouth, Hants, England.

Ci   Ciba (A.R.L.) Ltd., Duxford, Cambridge, England.

Co_1 Corblin et Cie, 78–80 Boulevard St. Marcel, Paris 5e, France.

British agent: C. T. (London) Ltd., 27 Ashley Place, London, W.1., England.

Co_2 The Cosmodyne Corp, 2920 Columbia St., Torrance, California 90509, USA.

Co_3 Compoflex Co. Ltd., 306 Burbury Street, Lozells, Birmingham 19, England.

Co_4 Consolidated Electrodynamics Corp., 1500 Shamrock, Monrovia, California 91017, USA.

Cr_1 Crofts Ltd., Bradford, Yorks, England.

Cr_2 Cryogenics Inc., 5821 Seminary Road, Bailey's Crossroads, Alexandria, Virginia 22041, USA.

Cr_3 Cryenco – Cryogenic Engineering Company, 4955 Bannock Street, Denver, Colorado 82216, USA.

Cr_4 Cryocal Inc., PO Box 10176, Riviera Beach, Florida 33404, USA.

Cr_5 Cryogenic Research Co., PO Box 881, Boulder, Colorado 80302, USA.

Cr₆   Crawford Fitting Co., 29500 Solon Road, Solon, Ohio 44139, USA.

CS   CSF (UK) Ltd., 1 Cadogan Place, London, S.W1., England.

Ct   CTLS, 53 Bonn-Ippendorf, Gierolstrasse 15, W. Germany.
UK office: Cryotechnic Ltd. Apollo House, 56 New Bond Street, London, W.1., England.

Da₁   Datametrics Ltd., Upton Road, Watford, Herts, England.

Da₂   Davison Chemical Division, W. R. Grace & Co., 101 North Charles St., Baltimore, Maryland 21203, USA.

De₁   Degussa, Weissfrauenstrasse 9, Frankfurt am Main, W. Germany.
UK agent: Bush, Beach & Segner Bayley Ltd., Engineering Division, Marlow House, Lloyds Avenue, London, E.C.3., England.

De₂   R. & J. Dempster Ltd., Chapter Street, Newton Heath, Manchester 10, Lancs, England.

Di   Distillers Co. Ltd., Great Burgh, Epsom, Surrey, England.

Dr   Drayton Hydroflex Ltd., Chantry Works, West Drayton, Middlesex, England.
also: Drayton Controls Ltd., Horton Road Works, West Drayton, Middlesex, England.

Du   Dunlop Rubber Co. Ltd., Fort Dunlop, Birmingham, England.
also: Dunlop Semtex Ltd., Insulating Division, Empire Way, Wembley, Middlesex, England.

Ea   H.G. East & Co., 37A Oxford Road, Cowley, Oxford, England.

Ed   Edwards High Vacuum Ltd., Manor Royal, Crawley, Sussex, England.

El₁   Electronic Instruments Ltd., Richmond, Surrey, England. (Formerly part of the Cambridge Instrument Company).

El₂   Elliott-Process Automation Ltd., Blackwall Lane, Greenwich, London, S.E.10., England.

En₁   Engelhard Industries, 207 Grant Avenue, Newark, NJ 07029, USA.
UK branch: Engelhard Industries Ltd., Valley Road, Cinderford, Gloucestershire, England.

En₂   Enthoven Solders Ltd., Dominion Buildings, South Place, London, E.C.2., England.

Ex₁   Expanded Perlite Ltd., 175 Piccadilly, London, W.1., England.

Ex₂   Expanded Rubber & Plastics Ltd., Mitcham Road, Croydon, Surrey, England.

Fi₁   Fischer & Porter Ltd., Warminster, Pennsylvania, USA.
UK branch: Salterbeck Trading Estate, Workington, Cumberland, England.

Fi₂   Fielden Electronics Ltd., Wythenshawe, Manchester, England.

Fi₃   Fisher Governor Co., PO Box 190, Marshalltown, Iowa 50158, USA.
UK branch: Airport Works, Maidstone Road, Rochester, Kent, England.

Fi_4 — Fine Tubes Ltd., Estover Works, Crownhill, Plymouth, Devon, England.

Fo_1 — Foxboro-Yoxall Ltd., Redhill, Surrey, England.

Fo_2 — Fowler & Polglaze Ltd., 8–9 Stephen Buildings, Gresse St., London, W.1., England.

Ga_1 — Gardner Cryogenics Corp., 2136 City Line Road, Bethlehem, Penna 18017, USA.

European Head Office: 20 Chaussée d'Houtem Vilorde, Brussels, Belgium.

UK office: 34 Hayseech, Cradley Heath, Warley, Worcs, England.

Ga_2 — Garrett Corporation, 9851 Sepulveda Boulevard, Los Angeles, California 90009, USA.

Ge_1 — General Electric Company, Schenectady, NY 12305, USA.

Ge_2 — General Engineering Co. (Radcliffe) Ltd., Yorkshire, England.

Gl — Gloucester Controls Ltd., Eastern Avenue, Gloucester, England.

Gr_1 — Greiner Scientific Corp., 20–26 N. Moore St., New York, NY 10013, USA.

Gr_2 — Gresham Lion Electronics Ltd., Twickenham Road, Feltham, Middlesex, England.

Gu — Guide Bridge Rubber Co., Vulcan Mill, Butcher Lane, Bury, Lancs, England.

Ha_1 — Harvey Control Company, 2880 Chattleton Lane, San Pablo, California 94806, USA

Ha_2 — Hale Hamilton (Valves) Ltd., Frays Mill Works, Cowley Road, Uxbridge, Middlesex, England.

Ha_3 — Harvey Fabrication Ltd., Woolwich Road, London, S.E.7, England.

He_1 — Henry Righton & Co. Ltd., 70 Pentonville Road, London, N.1, England.

He_2 — Henry Wiggin & Co. Ltd., Holmer Road, Hereford, England.

Hi_1 — Hird-Brown Ltd., Bolton, Lancashire, England.

Hi_2 — Hick, Hargreaves & Co. Ltd., Soho Ironworks, Bolton, Lancs, England.

Ho_1 — Hofman Division, Minnesota Valley Engineering Inc., New Prague, Minnesota 56071, USA.

Ho_2 — Honeywell Inc., Philadelphia Div., 1100 Virginia Drive, Fort Washington, Penna 19034, USA.

UK office: Honeywell House, Great West Road, Brentford, Middlesex, England.

Ho_3 — Hoke Mfg. Co., 1 Tenakill Park, Cresskill, NJ 07626, USA.

UK agent: George Meller Ltd., 26 Hallam Street, London, WIN 6LB, England.

Ho_4 — Honeywill-Atlas Ltd., Devonshire House, Mayfair Place, Piccadilly, London, W.1, England.

Ho_5 — F. J. Hone & Co. Ltd., 19 Eldon Park, London, S.E. 25, England.

Hy_1 Hymatic Engineering Co. Ltd., Redditch, Worcs, England.

Hy_2 Hygrodynamics Inc., 949 Selim Road., Silver Spring, Md.–Washington, D.C., USA.

In_1 Inertia Switches Ltd., 123 London Road, Camberley, Surrey, England.

In_2 International Gas Detectors Ltd., Great Wilson Street, Leeds, 11, England.

In_3 Intersonde Ltd., Queen Mary's Avenue, Watford, Herts, England.

In_4 International Research and Development, Fossway, Newcastle-upon-Tyne, 6, England.

IV I.V. Pressure Controllers Ltd., 683 London Road, Isleworth, Middlesex, England.

Ja Janis Research Company Inc., 22 Spencer Street, Stoneham, Mass. 02180, USA.

Je Jencons (Scientific) Ltd., Mark Road, Hemel Hempstead, Herts, England.

Jo_1 James Jobling & Co. Ltd., Sunderland, Co. Durham, England.

Jo_2 John Thompson (Pressure Vessel Division) Ltd., Ettingshall, Wolverhampton, England.

Jo_3 John Moncrieff & Co., Perth, Scotland.

Jo_4 Johnson Matthey Metals Ltd., 81 Hatton Garden, London, E.C.1, England.

KD KDG Instruments Ltd., Manor Royal, Crawley, Sussex, England.
Agent: A. D. Wood Ltd., Service House, 1 Lansdowne Road, Tottenham, London, N.17, England.

Ke Kent Instruments Ltd., Biscot Rd., Luton, Bedfordshire, England.

Ki Kinney Vacuum Div., 3529 Washington, Boston, Mass 02130, USA.
UK: see Ge_2

Le Leybold-Heraeus Ltd., Blackwall Lane, London, S.E.10, England.

Li Gesellschaft für Lindes Eismaschinen AG., 8021 Hollriegelskreuth bei München, W. Germany.

Lu Luchard et Cie, 137 à 161 Avenue Georges Clemenceau, Nanterre (Seine) Boîte Postale No.31, France.

Ma_1 Malaker Laboratories Inc., West Main Street, High Bridge, NJ 08829, USA.

Ma_2 Mason Renshaw Industries, Box 445, Carpinteria, California 93013, USA.

Ma_3 Magnion Inc., Cryo. Division, 144 Middlesex Turnpike, Burlington, Maryland 01803, USA.

Ma_4 C. S. Madan & Co. Ltd., Vortex Works, Altrincham, Cheshire, England.

Ma_5 Marston Excelsior Ltd., Fordhouses, Wolverhampton, England.

Me Meterflow Ltd., Royston Road, Baldock, Herts, England.

Mi	Mine Safety Appliances Co. Ltd., Marshgate Trading Estate, Taplow Road, Taplow, Bucks, England.
Mo	Monsanto Chemical Co., 800 N. Lindberg Avenue, St. Louis, Missouri 63166, USA. UK office: Monsanto House, Victoria Street, London, S.W.1, England.
Ne	Negretti & Zambra Ltd., Stocklake, Aylesbury, Bucks, England.
No	North American Philips Co. Inc., Cryogenic Division, PO Box 2200, Ashton, R.l. 02864, USA.
Nu	Nupro Company, 15635 Saranac Road, Cleveland, Ohio 44110, USA.
Ox	Oxford Instrument Co. Ltd., Osney Mead, Oxford, England.
Pa	Parkinson & Cowan Ltd., Talbot Road, Stretford, Manchester, England.
Pe	Peter Spence Ltd., Widnes, Lancs, England.*
Ph	N. V. Philips Gloeilampenfabrieken, Eindhoven, Holland. UK subsidiary: Pye-Unicam Ltd., York Street, Cambridge, England.
Pl	G. A. Platon Ltd., Wella Road, Basingstoke, Hants, England.
Po_1	Power Flexible Tubing Co. Ltd., Derby Works, Vale Road, London, N.4, England.
Po_2	Power-Gas Corporation, PO Box 21, Stockton on Tees, England.
Pr_2	Pressure Control Ltd., Davis Road, Chessington, Surrey, England.
Py	Pye-Unicam Ltd., York Street, Cambridge,
Qu	USA-Qualitrol Corp., 1387 Ford and Fairport Road, Fairport, NY 14450, USA. UK branch: Qualitrol Instruments Ltd., 97–107 Uxbridge Road, London, W.5, England.
Ra	Radiation Research Corp., 1150 Shames Drive, Westbury, Long Island, NY, USA.
Re	Reavell & Co., Ipswich, Suffolk, England.
Ro_1	Ronan & Kunzle Inc., 1225 S. Kalamazoo Ave., Marshall, Minnesota 49068, USA.
Ro_2	Rotameter Manufacturing Co. Ltd., 330 Purley Way, Croydon, Surrey, England.
Ro_3	H. Rollet & Co. Ltd., 36 Rosebery Avenue, London, E.C.1, England.
Ro_4	Rosemount Engineering Co. Ltd., Durban Road, Bognor Regis, Sussex, England.
RT	R.T.Z. Metals Ltd., York House, Empire Way, Wembley, Middlesex, England.
Ry	Ryan Industries Inc., 888 East 70th Street, Cleveland, Ohio 44103, USA.

*Now Laporte Industries Ltd., Moorfield Road, Widnes, Lancs., England.

Sa	Saunders Valve Company Ltd., Cwmbran, Mon, England.
Sc	Scientific Supplies Co. Ltd., Scientific House, Vine Hill, London, E.C.1, England.
Se_1	Servomex Controls Ltd., Crowborough, Sussex, England.
Se_2	S.E. Laboratories (Engineering) Ltd., North Feltham Trading Estate, Feltham, Middlesex, England.
Sh_1	Sherwood Overseas Corp, 65 E. 55 Street, New York, NY 10022, USA.
	UK Agent: Gatwick Electrode Co., Crawley, Sussex, England.
Sh_2	Shaw Moisture Meters, Rawson Road, Westgate, Bradford, Yorks, England.
Sh_3	Shell International Petroleum Co. Ltd., Plastics & Rubbers Division, Shell Centre, London, S.E.1, England.
Si_1	Silbrico Corp., 6300 River Road, Hodgkins, Illinois 60527, USA.
Si_2	Silica Gel Ltd., 62 Shaftesbury Avenue, London, W.1, England.
Sm	Smith's Industrial Instruments Ltd., Chronos Works, North Circular Road, London, N.W.2, England.
Sp	Spembley Technical Products Ltd., Trinity Trading Estate, Sittingbourne, Kent, England.
St	Stowe Laboratories Inc., Kane Industrial Drive, Hudson, Mass 01749 USA.
	UK agent: Guest International Ltd., Nicholas House, Brigstock Road, Thornton Heath, Surrey, England.
Su_1	Superior Air Products Co., 132 Malvern Street, Newark, NJ 07105, USA.
Su_2	Sulfrian Cryogenics, 391 East Inman Ave., Rahway, NJ 07065, USA.
Su_3	Sulzer Bros. Ltd., Winterthur, Switzerland.
Ta	Tangyes Ltd., Cornwall Works, Smethwick, Birmingham, England.
Te_1	Texas Instruments Inc., 3609 Buffalo Speedway, Houston, Texas 77036, USA. and Texas Instruments Ltd., Manton Lane, Bedford, England.
Te_2	Teddington Aircraft Controls Ltd., Bellows Division, Teilo Works,. Pontardulais, Swansea, Wales.
Ti	Tilbury Cryo-Equipment, 29 Kingston Road, Oxford, England.
Tr	TBT (see Ai_1).
Tu	Tube Sales (U.K.) Ltd., West Bay Road, Southampton, England.
Un_1	Union Carbide Corp., Linde Division, 270 Park Avenue, New York, 10017, USA.
	UK branch: 8 Grafton Street, London W.1, England.
Un_2	United States Steel, 1401 Arch Street, Philadelphia, Penna 19105, USA.

Un₃   United States Dynamics, 10 Denham Street, Newton Highlands 61, Mass., USA.
Va   Vacuum Barrier Corp., 4 Barton Lane, Woburn, Mass. 01801, USA.
Vi   Visco Ltd., Stafford Road, Croydon, Surrey, England.
Wa   Walker, Crossweller & Co. Ltd., Whaddon Works, Cheltenham, Glos., England.
We   Wesley Coe Ltd., Scotland Road, Cambridge, England.
Wh₁   Whittaker Nuclear Metals Division, Main Street, West Concord, Mass. 01781, USA.
        European agent: Eskenazi S. A., 24 Rue Joseph Girard, 1227 Carouge-Genève, Switzerland.
Wh₂   Whitey Research Tool Co., 5525 Marshall Street, Oakland 8, California, USA.
Wi   Williams & James Ltd., Chequers Bridge, Gloucester, England.
Wo   A. D. Wood (London) Ltd., Service House 1, Lansdowne Road, Tottenham, London, N.17, England.
Yo   Yorkshire Imperial Metals Ltd., PO Box 166, Leeds, England.

Index